162-

Chests of Drawers

CHESTS OF DRAWERS

Outstanding Projects from America's Best Craftsmen

WITH PLANS AND COMPLETE INSTRUCTIONS
FOR BUILDING 7 CLASSIC CHESTS OF DRAWERS

BILL HYLTON

The Taunton Press

 The Taunton Press
Inspiration for hands-on living™

The Taunton Press, Inc., 63 South Main Street,
PO Box 5506, Newtown, CT 06470-5506
e-mail: tp@taunton.com

Distributed by Publishers Group West

INTERIOR DESIGNER: Lori Wendin
LAYOUT ARTISTS: Cathy Cassidy, Lori Wendin
COVER PHOTOS: Seth Janofsky, Bob Gregson
ILLUSTRATOR: Melanie Powell

Library of Congress Cataloging-in-Publication Data
Hylton, William H.
   Chests of drawers : outstanding projects from America's best craftsmen :
with plans and complete instructions for building 7 classic chests of drawers /
Bill Hylton.
     p. cm.
  ISBN 1-56158-422-3
   1. Chests. 2. Furniture making. I. Title.

TT197 .H9496 2002
684.1'6--dc21

                       2001057377

About Your Safety
Working with wood is inherently dangerous. Using hand or power tools
improperly or ignoring safety practices can lead to permanent injury or even
death. Don't try to perform operations you learn about here (or elsewhere)
unless you're certain they are safe for you. If something about an operation
doesn't feel right, don't do it. Look for another way. We want you to enjoy the
craft, so please keep safety foremost in your mind whenever you're in the shop.

Printed in the United States of America
10 9 8 7 6 5 4 3 2 1

To babies Helen and Claire, twins, my first grandchildren,
who developed in utero while this book developed in computero.
Here's a manual to accompany your Double Dresser.

ACKNOWLEDGMENTS

I have been writing books for 30 years, and this is the first book in which I have the opportunity to acknowledge the assistance of others by name. This is a great thing.

I am indebted to Tom Clark, my editor, for giving me the opportunity to tackle this. But I am more indebted to him for his perseverance on my behalf and his extraordinary patience. Once we got started, I couldn't put him off, though Lord knows I tried. Having been in similar shoes, I *know* what grief was surely visited upon him for my pace. So I am especially grateful.

Thanks must go to the craftsmen—Glen Huey, Mark Edmundson, Harry Smith, Michael Seward, and Ken Burton—for making the chests featured in the book. Without these chests, no book. These craftsmen spent hours with me, either in person, on the telephone, or on-line, explaining how they did what they did. Michael and Ken set up photos as well. And Ken, a neighbor and former office-mate, shared ideas and know-how throughout the project.

One of the great things about this book-writing business is that you get to rub shoulders with some masterful furniture makers and can pick their brains about wood and tools and techniques and design. I learned from each of the contributing craftsmen, and I hope I was successful in conveying their know-how in the step-by-step instructions.

The photography in the book, with just a few exceptions, is the work of Donna Chiarelli, "my" photographer. I enjoy working with Donna—this is our third book. Propping and staging how-to photos can be tedious; but with Donna, getting the work done is always fun and the results are always exceptional. Besides that, my bulldog Matilde loves her.

Most important, I needed the love and support of my wife, Judi. A part of me whispers that "thanking the spouse" is a corny riff, but her contribution can't be ignored. For me, doing a book always consumes every waking hour and taints everything else that's going on. For more than 30 years, Judi's always been supportive and sympathetic.

CONTENTS

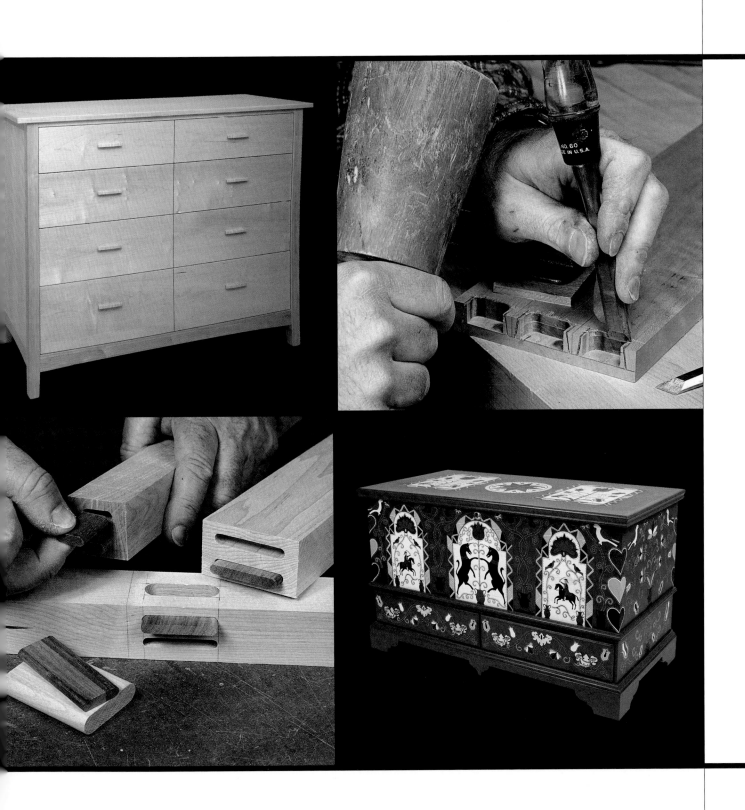

INTRODUCTION

A chest of drawers is a curious, paradoxical piece for the furniture maker. It is a major project, filled with challenge and opportunity. It provides the opportunity to show off your eye for design, your furniture engineering know-how, and your workmanship. Till it is done, you invest heavily, in terms of both time and materials. Perhaps even emotionally. The latter is especially true for the furniture maker who is making the chest of drawers for a loved one, maybe even for himself or herself.

A part of the paradox is the nature of the piece. There's usually a heavy aesthetic component to it. It has to look good. The right proportions are essential. You want to use the best materials, the most beautiful woods. You want it to reflect only your best—tight joinery, smooth surfaces, crisp edges.

And yet it's all about function, about *storage*. It's about socks and knickers, shirts and sweaters. Jeans. Belts. Handkerchiefs. Jewelry. Treasured knickknacks.

You want it to be beautiful, a showpiece; and yet, when it is all done, you are going to put it in your bedroom, the second most private room in the house (after the bathroom). And the typical visitor isn't going to see it.

Another aspect of the paradox is the interplay of the simple and the complex. When you strip it down to its elements, a chest of drawers is a big box that's filled with smaller boxes. Even a relative novice at woodworking has dealt with boxes. Simple project: sides, front, back, bottom. You are going to make a bunch of them, but each is simply sides, front, back, bottom.

But it does get complicated, because you don't really want to just pile up the small boxes inside the big one. You want to *arrange* the little boxes. You want to *graduate* them. Big boxes for your big duds (bulky sweaters, bib overalls) and smaller ones for your more delicate ones (ties). You want to be able to get into any one of those boxes without disturbing all the others.

All these things make the chest of drawers such a great project for any woodworker. There's the concrete functional analysis at the outset, the abstract creative design work, the problem-solving evaluation and planning, the thrilling hunt for and gathering of suitable materials. And the hours and hours of shop time.

My hope is that this book will help moderate the complications and get you into the shop a little sooner, a little better prepared. It's got two general chapters surveying case and drawer construction, explaining how these components are engineered and built. Options are presented and assessed for those who want guidance for creating and building an original design.

Following those chapters are seven excellent chest of drawer projects. You can fan through the pages and see what they look like. The same quick trip through the pages can give you an idea of how each is constructed—just look at the drawings and the how-to photographs.

The text of each chapter provides a rational sequence for building the piece, usually reflecting the approach used by the craftsman who designed and built the original.

The menu is diverse, including both traditional and contemporary styles and constructions. The first chests you come to are fairly small, but the tall chest that concludes the book is a huge one. None, frankly, is *really* simple, simple in the way a bookcase or a coffee table project can be. But they're as simple as a case piece with two to five drawers can be. Just work the plan. It will turn out to be a lot less complicated than you think.

CHEST-BUILDING BASICS

BUILDING A CHEST OF DRAWERS is—at its very base—building boxes. So what is it about such a project that's so off-putting for so many woodworkers? Perhaps it is the scale and the presumed intricacy of the construction, for the trepidation goes well beyond the aesthetics.

Every furniture maker, whether aspiring or a practicing, has criteria he or she uses to judge a piece of furniture. For me, a successful chest of drawers must be attractive, of course. But it has to be sturdy and utilitarian too. The drawers need to be strong and rigid, but not too heavy. The case has to be square, rigid, strong, and sturdy.

As a woodworker, I'm judgmental about the construction. That has to be rational. You want to make it as simple as possible, without giving up strength and durability and good looks.

Good examples of chests of drawers that meet these criteria are the seven projects in this book. But before we get to those projects, let's survey some chest of drawer construction methods.

If you intend to grow as a woodworker, you need to understand the appropriate joinery and the tested-and-proven construction approaches. If nothing else, this know-how will help you understand the choices made by each of the craftsmen represented in the book. You'll see that there's a logical progression in constructing a chest of drawers. Build the big

box first; then make the smaller ones to fit their places.

It is simple and complicated at the same time. Just a few parts, and such a lot of components.

PARTS OF A CHEST OF DRAWERS

The typical chest of drawers comprises a case, the drawers (of course), some sort of base or feet, and often trim or moldings. These are the chest's parts, and each plays a unique role in the structure. Yet each of these elements, perhaps with the exception of the molding, is itself an assemblage of many components

The case

The case is the big box. (A lot of people call this assembly a carcase, but that's too much like roadkill for me. I prefer the less visceral term.) Reduced to its most elemental form, the case is composed of sides, top, bottom, and back. Its job is to organize and hold all the drawers.

To do that job, it has to be divided up so there's a compartment for each drawer. The typical case has drawer dividers and runners and sometimes drawer guides and kickers. Now and then, you come across a chest of drawers that uses manufactured runners to support the drawers, and such a chest usually has no dividers.

Chests of drawers can be made by using traditional construction or by employing some modern strategies. The Double Dresser (p. 116) uses mechanical drawer slides instead of conventional drawer dividers (also called rails) and drawer runners.

The *drawer dividers* are horizontal members that extend from side to side, dividing the big box into a series of full-width compartments for drawers. These dividers are sometimes called drawer rails. *Vertical drawer dividers* split full-width compartments so the case can have two or more drawers that are narrower than the width of the case.

The *runners* are what really support the drawers. These typically extend from the back edge of the divider to the rear of the case, and they are mounted to the case side. Where there are drawers that are less than full width, you need an intermediate runner—one that will support two drawers—and the intermediate runner dictates the use of a back rail. *Guides* prevent side play as a drawer is opened and closed. A *kicker* prevents the drawer from tipping down as it is opened.

The base

Do you ever want your chest of drawers to rest on a floor, with the bottom drawer just ¾ in. or so clear of it? I can't think you ever would, so some sort of feet or base is essential.

Chest bases can range from a simple bracket foot to a more elaborate framework, shown here on the Queen Anne Chest on Frame (p. 166).

The options here are numerous, ranging from integral feet to a complete, separate leg-and-apron stand. The integral feet are simple, often being extensions of the corner posts. Separate feet can involve added-on posts, blocking to support the posts, and facings to hide the posts.

The trim

Moldings are style setters. These concave and convex surfaces, collectively called profiles, have undeniable impact on the appearance of a chest of drawers.

They ease transitions between parts and surfaces, divide up large open areas, and refine edges, corners, and other borders. The size and shape of a molding can enhance (or upset) the visual balance or scale of a piece. The moldings used across the top and bottom of a chest are usually the most prominent. These moldings separate the piece from the room. A large case often gets robust moldings, whereas a smaller chest needs more subtle moldings.

When a chest consists of two cases, a waist molding provides a visual transition from one to the other and simultaneously provides a mechanism to link the two cases structurally. The molding generally is attached to the lower case in such a way that it creates a lip inside of which the upper case just fits. The molding traps and aligns the upper case, and gravity holds it in place.

DESIGNING A CHEST OF DRAWERS

In designing a chest of drawers, your goal may be artistic. Certainly, some chests look handsome and may, in fact, be works of art. But don't lose sight of the fact that their function is humble: *storage*.

The baseline form is the lidded chest. Although it isn't a chest of drawers, it gives us a chance to consider the lowest level of storage organization. With this kind of chest, the belongings are organized only if you are.

My wife, Judi, has a big cedar-lined blanket chest that is a repository for seldom-used heavy blankets, a few quilts sewn by her mother and grandmother, bed linens for the guest bedroom, clothes from our kids' pasts that are too treasured to part with, dolls and toys from her own childhood, and even a box with a thick rope of her own hair, shorn . . . hmmm . . . 25 years ago. Locating any one item is an archaeological excavation. She lifts off the top layer and sets it aside. Is it exposed now? Layer by layer, she removes the chest's contents until she finds what she's looking for. In a hurry? She'll rummage down without removing anything, locate her object, and just pull it out, tumbling everything that was on top of it. Good-bye organization.

This is just the sort of storage that the chest of drawers was conceived to eliminate. The blankets and sheets can be separated from each other and from the clothes. Depending on the number of drawers and their sizes, Judi may be able to put her rope of hair in its own drawer.

Drawers resolve many storage problems. The challenge is to size both the drawers and the chest to provide a workable level of organization while making the piece physically easy to use.

Here are some commonsense rules:

- Big drawers at the bottom; small drawers at the top.
- No drawer too big to open and close (though modern hardware has raised the dimensions threshold here).
- No drawer too high to see into.
- And not so many drawers that you can't remember what—generally speaking—is stored in each.

To make it manageable for the widest range of people, the largest drawer should not be more than 12 in. high, 24 in. deep, and 48 in. wide. Such a drawer should be at the bottom of the chest. Not only will it be easier to open and close but it will look better.

It's a good idea to scale a chest of drawers for the person who will use it most. There's a lot of tradition in this.

A form like the chest on chest is generally considered "the man's chest." With heights ranging up to 84 in. and widths up to 48 in., these can be colossal pieces that press the

ergonomic envelope. At least a pair or trio of drawers are above eye level for all but the tallest people. Its full-width drawers can be awkward for a small person to deal with.

A more manageable form, often regarded as being for "the woman," is the low, wide chest of drawers, usually called a dresser. Its overall height of 29 in. to 34 in. puts all the drawers below adult eye level. Positioning drawers in side-by-side stacks keeps their widths manageable, even in a piece 72 in. long.

CONSTRUCTING A CASE

Building a case can seem so simple. You join two sides to a top and bottom to form a box. You build in an internal framework to support drawers, and you add a back.

Ah, it should be so simple. To me, there are three basic ways to construct a case, none of which is complete without the internal structure needed to support drawers. But let's focus first on the basic box.

If you use solid wood—as opposed to manufactured wood products like plywood and medium-density fiberboard (MDF)—you want to focus on using a limited number of constructions and a narrow selection of joints. Wood continually expands and contracts with changes in temperature and humidity. You can't prevent it, so you have to deal with it.

If you use manufactured wood products—which are stable and which don't expand and contract—you have a slightly different assortment of construction and joint options.

A compromise approach, frame-and-panel construction, uses all solid wood to produce stable case components. It is more work, but to some eyes it is more attractive. It definitely circumvents wood-movement problems.

Let's look at each type of construction.

The solid-wood case

The fundamentals of solid-wood case construction are shown in the drawing on p. 8. The big issue is wood movement. Basically, wood expands and contracts across the grain in response to seasonal humidity changes. For all practical purposes, however, wood is

SOLID-WOOD CASE JOINERY

If you build the case with solid-wood sides, it's best to also use solid wood for the top and bottom. Orient the grain consistently, and all the parts will move in tandem. If you use an open frame for the top or bottom, you must use joinery that won't hamper the seasonal expansion and contraction of the sides.

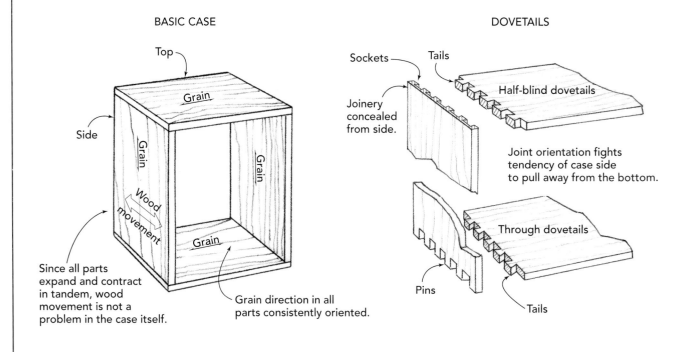

BASIC CASE

Top

Grain

Side

Grain

Grain

Grain

Grain

Wood movement

Since all parts expand and contract in tandem, wood movement is not a problem in the case itself.

Grain direction in all parts consistently oriented.

DOVETAILS

Sockets

Tails

Half-blind dovetails

Joinery concealed from side.

Joint orientation fights tendency of case side to pull away from the bottom.

Through dovetails

Pins

Tails

ALTERNATIVE JOINERY

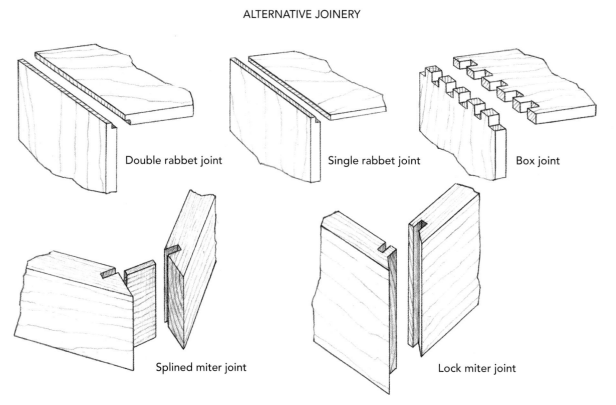

Double rabbet joint

Single rabbet joint

Box joint

Splined miter joint

Lock miter joint

dimensionally stable along the grain. Because of this difference in movement, grain orientation is a key consideration in solid-wood construction. All the components—the sides, top, and bottom—need to have the wood grain going in the same direction. When the wood expands or contracts, all the components move together in the same direction.

For solid-wood construction, therefore, you'll want to use a solid top and bottom (as opposed to open frames) to match the sides. The seasonal movement of the solid-wood top and bottom will match that of the sides. But an open framework with long-grain pieces attached to the sides will conflict with the movement of the sides.

The traditional joint for linking the sides, top, and bottom together is the dovetail; both through and half-blind dovetails are used. A dovetail joint that's accurately cut and well fitted offers a mechanical connection. Apply glue to its many interlocking surfaces and you have a superior joint.

The dovetail usually is viewed as a sign of quality craftsmanship. A woodworker may expose them in a chest, just so you can see they were used. Two hundred years ago, dovetails would be concealed by a molding. Or the half-blind version would be used so a you'd see only the face grain. Moreover, dovetails are often avoided by woodworkers who haven't mastered the skill of hand cutting them and don't want to machine cut them.

Alternatives to dovetails abound, as the drawing indicates. Box joints, a sort of machine-cut equivalent of the dovetail, are effective though tedious to cut. Rabbet joints, whether single or double, are simple to cut yet quite effective. Miters of various sorts are aesthetically pleasing, since they leave no end grain exposed.

The benefits of solid-wood construction are straightforward. The basic box of sides, top, and bottom is just that: four parts. When you orient the boards properly and assemble them with appropriate joinery, wood movement isn't a problem. And those gorgeous (and expensive), highly figured boards can dominate.

Dovetail construction is one of the strongest ways to create the casework for a chest of drawers. However, it is time-consuming and limited to solid-wood construction.

The sheet-goods case

Plywood and MDF are good materials options. Oh, some of you may cringe at that. These are materials for kitchen cabinets and built-ins, not for hand-crafted furniture like chests of drawers. But a vast array of hardwood veneers on plywood or MDF substrates are available, including exotics.

Sheet goods have a lot of positives. You don't have to dress the stock. You don't have to glue up boards to create 18-in.-wide panels. You don't have to agonize over matching the color and figure of the boards in a glue-up. Concerns about cross-grain constructions are out. Just glue those runners or that molding across the case side. Lay out your parts on the 4-ft. by 8-ft. sheet, make a few careful saw cuts, and you're ready for joinery cuts and assembly. When time is dear, sheet goods can be sweet.

SHEET-GOODS CASE JOINERY

Plywood (and MDF) can mimic solid wood in appearance, but these man-made sheet goods are very stable. You don't really need to account for wood movement in a case constructed primarily of plywood or MDF. While you can't use dovetails with these materials, there are a variety joints you can use—most of which can be cut quickly.

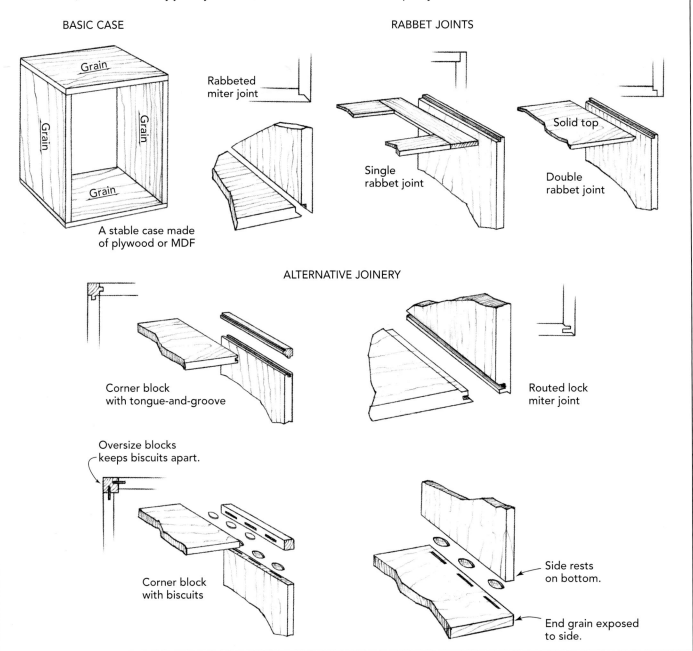

BASIC CASE

Grain
Grain
Grain
Grain

A stable case made of plywood or MDF

RABBET JOINTS

Rabbeted miter joint

Single rabbet joint

Solid top

Double rabbet joint

ALTERNATIVE JOINERY

Corner block with tongue-and-groove

Routed lock miter joint

Oversize blocks keeps biscuits apart.

Corner block with biscuits

Side rests on bottom.

End grain exposed to side.

The edges are ugly, of course, and need to be concealed. Glued-on ⅛-in.-thick strips of the appropriate hardwood will do.

The assortment of joinery that works is different from that for solid wood. No dovetails, no box joints. You can still use rabbets and various reinforced miters, however. In addition, you can use biscuit joints and corner block joints. The basics are shown in the drawing above.

Because plywood is stable, you can use an open frame in place of the top or bottom (or both).

The frame-and-panel case

Frame-and-panel construction and its close kin post-and-panel construction were surely contrived hundreds of years ago as a means of dealing with wood's instability. Today, it's the appearance that grabs people.

The fundamental point of frame-and-panel construction is to create a stable unit to use in place of an unstable board (see the drawing on p. 12). The change in dimension in a narrow board is going to be much less than that in a wide one. So you use four narrow strips of wood to form a frame; and, effectively, that frame will be stable. The rails and stiles making up the frame don't change in length, only in width and thickness.

Then there's the opening defined by the frame. A board built into the frame may be quite wide; and because it is wide, it's going to move quite a bit. To compensate, the panel is captured in a groove that keeps it in place whether it's shrunk to its minimum width or expanded to its maximum. While the panel shrinks and swells, the frame does not.

One of the most common frame-and-panel assembly approaches uses traditional mortise and tenon joinery, with the panel housed in grooves. A more contemporary method uses machine-cut cope-and-stick joinery.

In production terms, the cope-and-stick approach is probably faster. To cut the joinery, you use a matched pair of cutters in a table-mounted router or a shaper. The ends of the rails are machined with the cope cutter, which produces a stub tenon and the negative of whatever edge profile is cut by the sticking cutter. Then, with the sticking cutter, you machine one edge of all the frame parts, stiles and rail alike, to profile the edge and to slot the edge for the panel and the stub tenons. The frame is assembled by gluing the stub tenons into the slots (and the surface of the

Frame-and-panel construction can be done in solid wood or combined with plywood, as shown here. Although the system was designed to accommodate seasonal wood movement for solid wood, the layering of the parts is attractive in its own right.

FRAME-AND-PANEL CONSTRUCTION

This proven construction approach produces stable components from solid wood, the expansion and contraction across a 2"- to 3"-wide board is negligible. The unstable element—the panel—is captured by a groove in the frame, which includes room for expansion. The panel can be raised or flat and can be cut from solid wood or sheet goods.

BASIC CONSTRUCTION

Section View

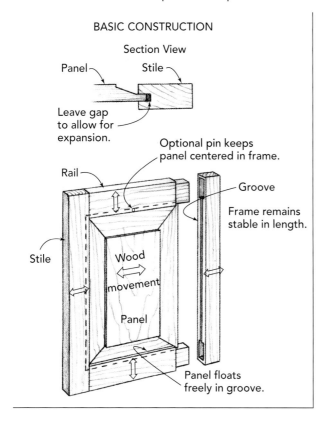

Panel

Stile

Leave gap to allow for expansion.

Optional pin keeps panel centered in frame.

Rail

Groove

Frame remains stable in length.

Stile

Wood movement

Panel

Panel floats freely in groove.

ALTERNATIVE CONSTRUCTION

Post-and-Panel Construction

Top rail

Panel groove in rail extends from mortise to mortise.

Panel can be raised or flat, solid wood or plywood.

Mortise-and-tenon joint

Post

Bottom rail

Extends past bottom rail to form foot.

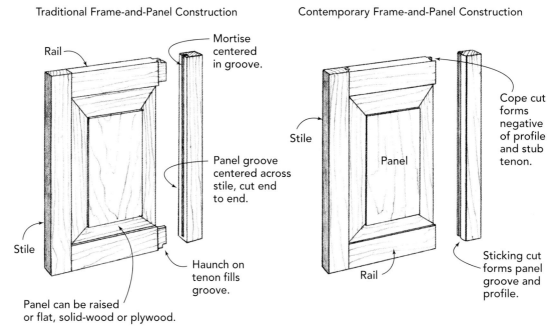

Traditional Frame-and-Panel Construction

Rail

Mortise centered in groove.

Panel groove centered across stile, cut end to end.

Stile

Haunch on tenon fills groove.

Panel can be raised or flat, solid-wood or plywood.

Contemporary Frame-and-Panel Construction

Stile

Panel

Rail

Cope cut forms negative of profile and stub tenon.

Sticking cut forms panel groove and profile.

cope to the profile). The panel floats, unglued, in grooves.

The assembly is plenty strong for cabinet doors, perhaps its most common use. In cases, the cope-and-stick unit is supported by web frames, back, top, and other structural elements.

Still, the well-established strength of the mortise-and-tenon joint may be more reassuring. And that may make the extra work this traditional approach requires worthwhile for you. You have four setups instead of two (mortising, tenoning, grooving, and profiling), and you have to hand-fit the frame at each joint to accommodate the profile.

In either approach, you can use any of a variety of panels: raised, flat, thick and proud, thin and recessed. The panel can be solid wood or plywood.

In the realm of casework, the post-and-panel variant of this construction is most common. The essentials of frame-and-panel construction still apply, except that the stiles are replaced with thicker stock posts, which extend beyond the bottom rail to serve as feet for the completed chest.

You can see already how much more work this approach takes than simply dressing stock, perhaps doing a couple of glue-ups, or cutting a couple or quartet of panels from a sheet of plywood. After you've dressed stock, cut parts and joints, and completed at least one assembly glue-up, you still must construct the case.

Combining frame-and-panel or post-and-panel assemblies with top and bottom elements to form a basic box isn't remarkably different from the case construction schemes discussed earlier.

Using solid-wood top or bottom panels with a post-and-panel side assembly is problematic, because that assembly is stable and the solid-wood panels are not. There isn't a good way to join the two elements. Instead, use web frames for the top and bottom; then apply a solid top over the top web frame. Of course, you can avoid using a web frame if the top or bottom solid panel is plywood or MDF.

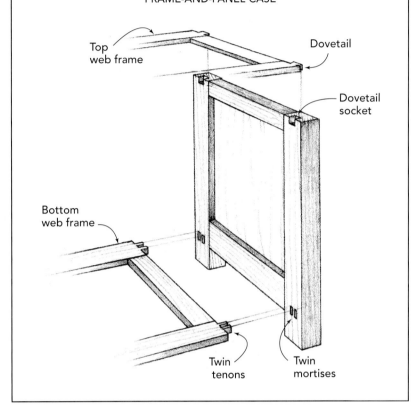

POST-AND-PANEL CASE CONSTRUCTION

Post-and-panel assemblies have the stability of plywood. Incorporating them into casework requires you to use a variety of joints, including mortises and tenons, dovetails, and rabbets.

FRAME-AND-PANEL CASE

Top web frame

Dovetail

Dovetail socket

Bottom web frame

Twin tenons

Twin mortises

The ideal joinery between a top frame and a post- or frame-and-panel side assembly is dovetails. For a post, cut one large tail on each end of the top-frame rails and a matching socket in the post top. For a stile, cut tails for half-blind dovetails on the top frame rail ends and sockets in the end-grain of the stiles. Don't carry the sockets into the side rails. In either situation, this joinery will keep the top end of the case from spreading open.

Mortise-and-tenon joinery also works, especially at the bottom and especially with posts. You simply can't cut a deep enough mortise in the stiles of a frame-and-panel assembly to get a strong joint; the stiles are just too thin. For the top of a post, several configurations of

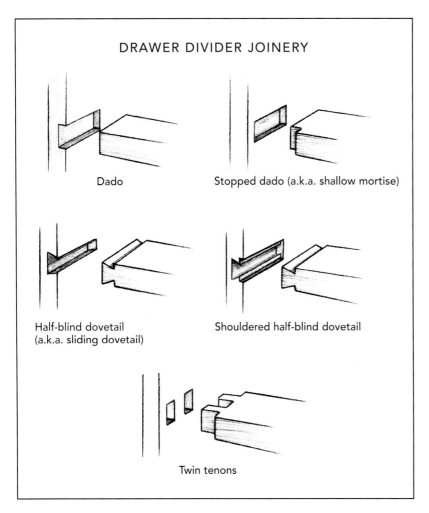

DRAWER DIVIDER JOINERY

Dado

Stopped dado (a.k.a. shallow mortise)

Half-blind dovetail
(a.k.a. sliding dovetail)

Shouldered half-blind dovetail

Twin tenons

is to divide the case, using drawer dividers or web frames, to create a separate compartment for each drawer. The runners for the drawers join the divider and the case sides or a back drawer rail.

Drawer dividers

Often called drawer rails, drawer dividers are horizontal members that extend from case side to case side. They separate one drawer opening from another. But more than that, they keep the case sides straight and parallel. As such, they need to be integral to the case's overall design and construction.

Drawer dividers can be joined to the sides in a variety of ways, as shown in the drawing at left.

The dado joints are easy to do and work for all constructions (solid-wood, plywood, and frame-and-panel). The dado cut can be through (exposed on the front edge); but most often, it is stopped. Where a stopped dado is used, the divider itself needs to be notched at the front corner. These joints present little resistance to tension stress; in theory, the case side could bow outward, pulling the joint apart.

The various types of dovetail joinery *do* give good resistance to tension stress, presenting a mechanical opposition to any tendency of the side to bow out. A router makes quick work of such cuts.

The twin tenon is well suited both to post-and-panel construction and plywood cases with solid-wood drawer dividers. In this sort of application, there's always a temptation to orient the tenon across the rail's width. But this creates a lot of long-grain-to-end-grain gluing surface, which isn't at all strong. You are much better off if you form two tenons oriented with the rail's thickness, thus maximizing the long-grain-to-long-grain gluing surface and creating a strong joint.

Mortises for the joint can be cut with consistency using a horizontal boring machine or a plunge router with a mortising jig. If you use the loose-tenon variant, you mortise both mating parts and make separate tenons.

mortises and tenons will work, but they aren't as good as the dovetail. At the bottom of a post, choose a configuration based on the rail orientation. If it is flat, use twin mortises and tenons; but if it is on edge, use a single large tenon, making it as deep into the post as possible.

Where the rails are flush with the inner faces of the posts (as in frame-and-panel assemblies), you can use rabbets and dadoes.

DIVIDING THE CASE

The methods of making and hanging drawers are covered in "Drawer-Building Basics" on p. 24. But the elements that support drawers in a case must be incorporated as the case itself is constructed. The traditional approach

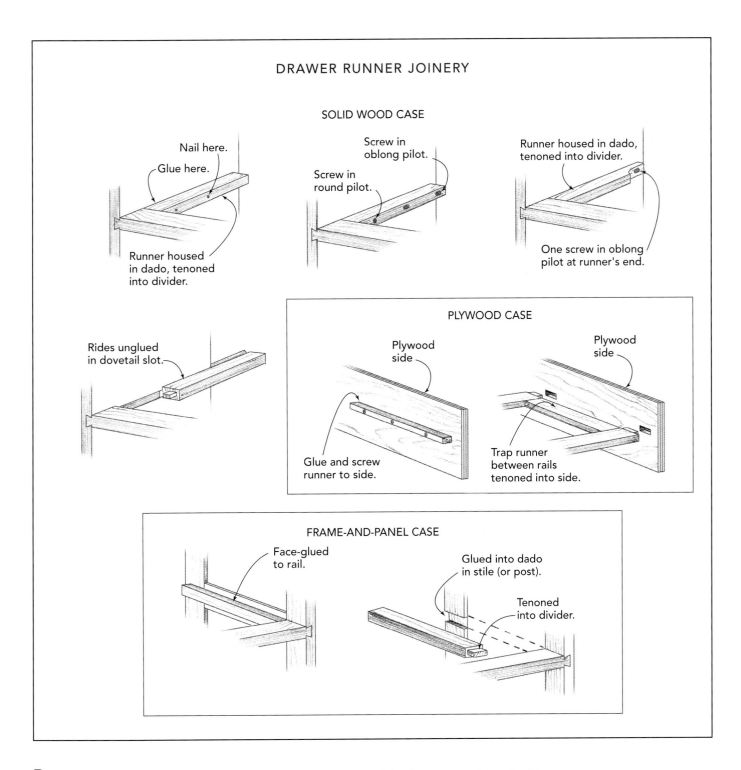

DRAWER RUNNER JOINERY

SOLID WOOD CASE

Nail here.

Glue here.

Runner housed in dado, tenoned into divider.

Screw in round pilot.

Screw in oblong pilot.

Runner housed in dado, tenoned into divider.

One screw in oblong pilot at runner's end.

Rides unglued in dovetail slot.

PLYWOOD CASE

Plywood side

Plywood side

Glue and screw runner to side.

Trap runner between rails tenoned into side.

FRAME-AND-PANEL CASE

Face-glued to rail.

Glued into dado in stile (or post).

Tenoned into divider.

Drawer runners

Runners support the drawer and allow it to slide in and out of the chest. The runner itself is usually a strip of the secondary wood, the same width as the drawer divider. It's linked to the divider, most often with a mortise-and-tenon joint (glued or not) and attached in some way to the case side.

In a solid-wood case, the runners are cross-grain to the side. Thus they have to be mounted so that the case side can expand and contract. Simply gluing a runner to the side can prompt

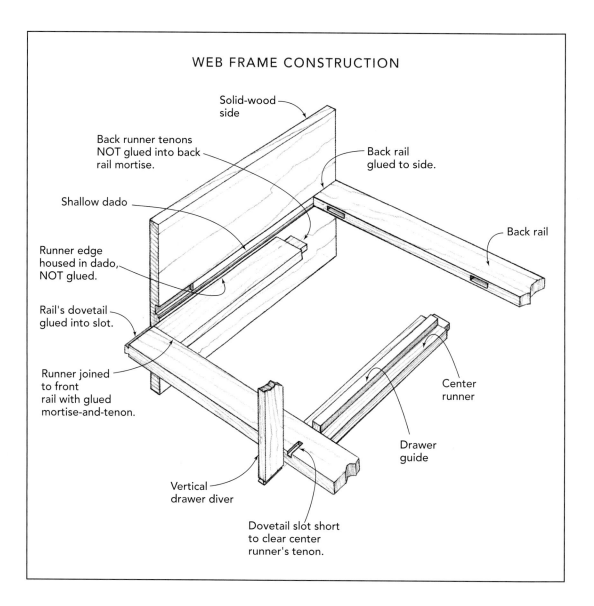

WEB FRAME CONSTRUCTION

Solid-wood side

Back runner tenons NOT glued into back rail mortise.

Shallow dado

Runner edge housed in dado, NOT glued.

Rail's dovetail glued into slot.

Runner joined to front rail with glued mortise-and-tenon.

Vertical drawer diver

Dovetail slot short to clear center runner's tenon.

Back rail glued to side.

Back rail

Center runner

Drawer guide

splitting and buckling of the side. For solid-wood cases, the runner is housed in a shallow dado. The dado joint takes the weight of the drawer, and the fastening options are all intended to hold the runner in the dado.

When the runner is being joined to a frame-and-panel assembly, wood movement is moot. The runner can be edge-glued to a rail in the side assembly, or it can be set into dadoes in the side assembly's stiles or posts and glued, so long as it isn't glued to the panel. In plywood construction, the runner can be glued into a shallow dado or glued and screwed directly to the plywood side.

Web frames

The system of drawer dividers and runners is often turned into a complete frame, with front and back rails and runners. The resulting frame, usually called a web frame, is essential when support is needed for drawers that are less than the full width of the case. The back rail supports the back end of a center runner and a guide. Web frames are also essential when the case lacks a solid top and bottom.

In a case with solid-wood sides, you might eschew glue in the traditional mortise-and-tenon joints between the runners and rails. When the rails are glued to the sides, the unglued joints with the runners allow the side to move; the tenons simply slip in and out of

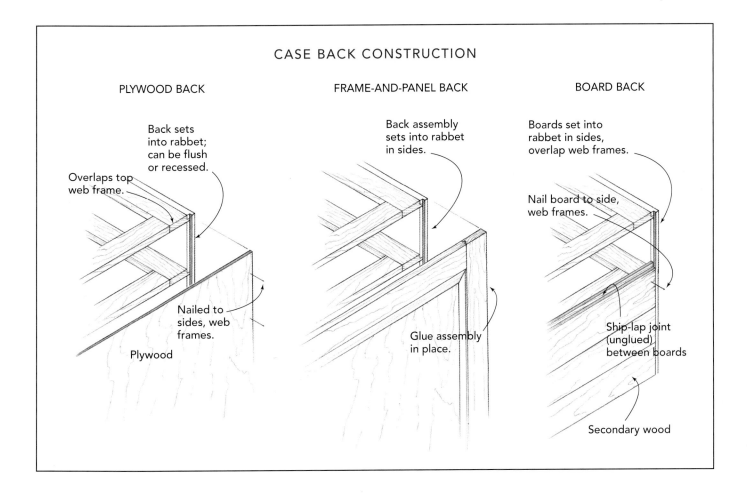

CASE BACK CONSTRUCTION

PLYWOOD BACK

Overlaps top web frame.

Back sets into rabbet; can be flush or recessed.

Nailed to sides, web frames.

Plywood

FRAME-AND-PANEL BACK

Back assembly sets into rabbet in sides.

Glue assembly in place.

BOARD BACK

Boards set into rabbet in sides, overlap web frames.

Nail board to side, web frames.

Ship-lap joint (unglued) between boards

Secondary wood

the mortises. In most cases, the runner should be slightly short—by about ⅛ in.—so the side can shrink, closing the gap between the runner and the back rail without pushing the rail out or cracking the side.

In a frame-and-panel case, the web frame can be glued up and then joined to the posts or rails with dado joints or twin tenons. The only caveat is to leave a solid-wood panel in the side assembly free to move; never glue the runner to the panel.

In a plywood case, it seems easiest to glue an assembled web frame into a dado cut across the side. But that sort of dado weakens the plywood. It's better if you join the rails and sides with twin tenons and glue the runner to the plywood between the rails.

A vertical divider can be joined to the drawer divider with the same array of joints used between it and the case sides: stopped dado, sliding dovetail, twin tenons. The verti-

cal divider typically is the same width as the horizontal divider. The runner behind it is wider than the others, simply because it supports two drawers and a guide. The guide is glued to the top face of the runner, directly behind and aligned with the vertical divider.

CLOSING IN THE BACK

The case for a chest of drawers is incomplete without a back. Besides closing in the case, the back braces it, keeping it from wracking. Occasionally, the back supports drawer runners or kickers.

Aesthetic demands vary, of course. In the typical chest of drawers, the back is strictly functional. What it looks like is of little or no consequence, because it's going to be against a wall. Occasionally though, a back has to present an attractive appearance (if only to satisfy the woodworker's vanity).

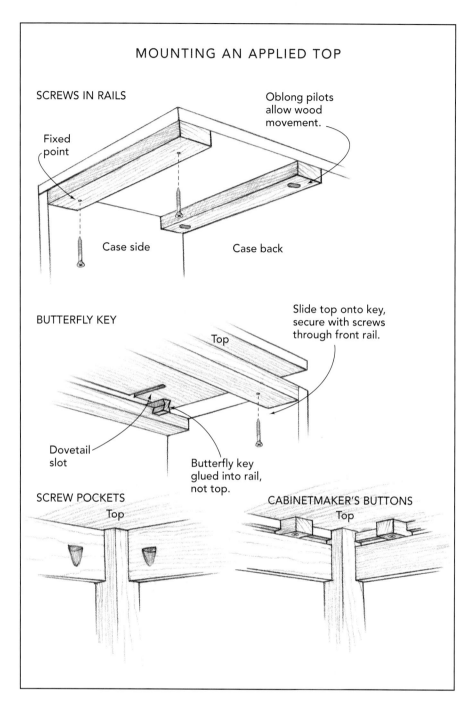

MOUNTING AN APPLIED TOP

SCREWS IN RAILS

Oblong pilots allow wood movement.

Fixed point

Case side

Case back

BUTTERFLY KEY

Slide top onto key, secure with screws through front rail.

Top

Dovetail slot

Butterfly key glued into rail, not top.

SCREW POCKETS

Top

CABINETMAKER'S BUTTONS

Top

Plywood

Plywood is today's solution. Strong, stable, easy to work with, economical, and light-weight, it is ideal for a case back. A hardwood plywood back can even be attractive if it is installed carefully. The most common installa-tion houses the plywood in a rabbet cut in the sides, top, and bottom. When the plywood is

slid into a groove in the sides and bottom before an overlaid top is mounted, the look can be very finished (see Double Dresser on p. 116).

Frame and panel

Most finished in appearance is the frame-and-panel back. Because it entails more labor and time to make than a plywood back, it is used primarily when the back of the chest will be exposed (for example, when the completed piece will be used in the center of a room).

Boards

The earliest chests of drawers had board backs. Oriented horizontally or vertically, the boards were traditionally recessed into rabbets in the sides and would often overlap the top and bottom, especially in a tall piece. In the most primitive chests, the boards are simply butted edge to edge. Commonly, though, the boards were joined with unglued shiplap or tongue-and-groove joints.

The board back is still with us, primarily in reproductions. The Queen Anne Chest on Frame (p. 166) and the Tall Chest (p. 188) use this type of back. The boards are cut from a secondary wood, ripped to random widths, and set edge to edge with unglued shiplap joinery. The boards are set into rabbets cut slightly deeper than their thickness and then nailed to the sides, top, and bottom with reproduction cut nails.

ATTACHING A TOP

A case needs a top. If the top isn't integral, it has to be attached. An attached top usually overhangs the case in front and on the sides, and it may have profiled edges.

A top is mounted to its case in the same way a tabletop is mounted to its frame. The drawing at left shows four good ways to do it. Each of these approaches holds the top firmly to the case while allowing it to move to the front and/or back with changing moisture conditions.

Usually, the top is fixed at the front (so it will always look the same in relation to the case's facade) and allowed to move to the back.

If you drive screws through the top web frame into the case top, use round pilots in the front rail and oblong pilots in the back rail. If you drive the screws through pockets in the sides, front, and back, you need to make the pilot holes in the sides and the back sufficiently oversize to permit movement of the top.

The butterfly key is an interesting construction technique invented by an eighteenth-century cabinetmaker. Butterfly keys are glued into slots in the back top rail, and matching slots are cut into the underside of the top. The top is then slid onto the keys—without glue—and attached to the front rail with screws.

CONSTRUCTING THE BASE

Seldom, if ever, does a chest of drawers rest directly on the floor. Instead, it has integral posts, a stand, a base, or added-on feet to support it. Countless variations and hybrid foot and base designs have been used.

Simple options

An obvious way to support a chest is to extend the sides past the lowest drawer rail, so the structure rests on the bottom edges. You can minimize the area in actual contact with the floor by cutting out the sides to form feet. This works, but its applicability is much pretty limited to primitive styles. More commonly, a baseboard is applied over the side extensions. The baseboard is usually cut out to form feet at the corners, and its top edge is shaped to provide a base molding. You can achieve the same appearance by using a separate base.

If the case is post-and-panel construction, it can be supported simply by extending the posts several inches past the bottom. The section from the bottom rails to the bottom end becomes the foot. It can be left square, it can be tapered, or it can be turned. This construction has been common from the Jacobean period of the seventeenth century through the Arts and Crafts movement of the late nineteenth century right up to the present day. Examples

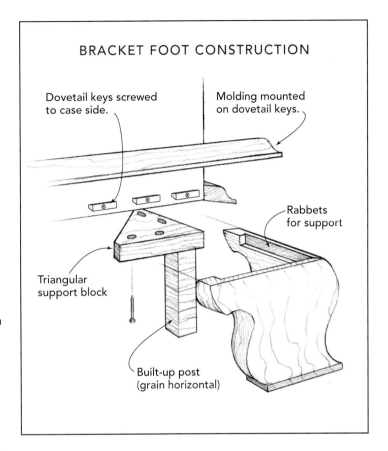

BRACKET FOOT CONSTRUCTION

Dovetail keys screwed to case side.

Molding mounted on dovetail keys.

Rabbets for support

Triangular support block

Built-up post (grain horizontal)

of this construction are the Contemporary Chest (p. 62) and the Double Dresser (p. 116).

The bracket foot

The bracket foot has been used on many furniture styles, from William and Mary to Queen Anne, country, Chippendale, and contemporary. While the traditional construction is not overly difficult, it does involve cross-grain glue joints, which can self-destruct.

In the traditional construction, a short vertical post is butted against the bottom of the case and takes its weight. Glue blocks of roughly the same size are glued to the bottom of the case and the inner faces of the bracket. The bracket is glued to the post, the case, and the glue blocks. The joint between the post and the bracket is cross-grain. A base molding glued to the case and to the bracket is cross-grain to the case sides.

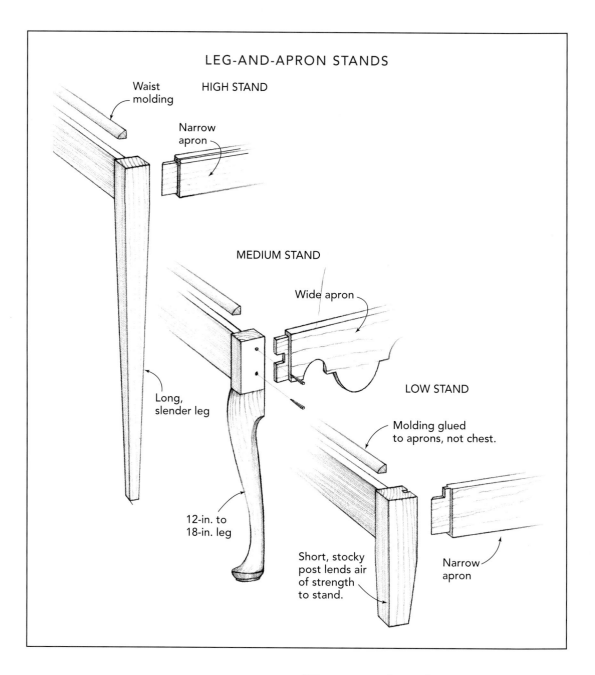

LEG-AND-APRON STANDS

HIGH STAND

Waist molding

Narrow apron

MEDIUM STAND

Wide apron

Long, slender leg

LOW STAND

12-in. to 18-in. leg

Molding glued to aprons, not chest.

Short, stocky post lends air of strength to stand.

Narrow apron

A better construction supplants the glue blocks with a triangular support block that's housed in stopped rabbets in the bracket. The post is built up, and its edge grain parallels the grain of the foot; it is less of a post, perhaps, than a glue block. It is not cross-grain though, and it will move together with the foot. The assembly is attached to the case with screws driven through the support block. The base molding is attached to the case sides with T-slots or dovetail keys, so the case can move without restriction.

The stand or base

The last option is to make a separate frame or leg-and-apron stand to hold the chest. Such a base can be constructed very sturdily, making a strong base. It is easy to attach it to a case in a way that accommodates wood movement. And it can be attractive as well.

Plinth base The plinth base is the most easily constructed separate base, but it isn't common on chests of drawers. According to the dictionary, a plinth is the lowest element in an architectural base. It is commonly perceived as being straight, flat, and unembellished. In furniture, a plinth traditionally is a little wider and deeper than the case it supports, with a molding easing the visual transition from base to cabinet. It is little more than a box frame, assembled with an appropriate case joint. A large frame might be reinforced with one or more cross members, and the structure can be stiffened with corner blocks. The base can be attached to the case bottom, with screws driven through corner blocks, pockets, cleats, or even cabinetmaker's buttons.

Bracket base A traditional plinth base is transformed into a bracket base by making cutouts to form "feet." This form is much more common on chests of drawers.

A very traditional bracket base is constructed with through dovetails, which produces a strong frame. But miter joints are commonly used these days at the exposed front corners. At the back, butt or rabbet joints hold the frame's side members to the back. Unembellished bracket feet cut from the secondary wood are sometimes used instead of a full-length back member.

Gluing a molding to the base circumvents problems that might arise from attaching it across a case side. Depending on the size of the molding and the thickness of the stock used for the base, a ledge for the case is formed. For broader support, cleats are glued to the inside of the base or let into rabbets in the base. The base is then screwed to the case bottom.

Leg-and-apron stand Build a very low framework for a table, but instead of mounting a tabletop, mount the chest of drawers on it. This leg-and-apron stand is the type of base used under the Queen Anne Chest on Frame (p. 166) and under the Triple Dresser (p. 138). The two chests are rather different, yet their bases, at least conceptually, are the same. See the drawing on the facing page. Bear in mind that the lengths of the legs may vary, but the joinery remains basically the same.

CONSTRUCTING AND MOUNTING MOLDINGS

Moldings, as I noted earlier, have an enormous impact on the appearance of a chest of drawers. Moldings make the transitions between the parts and surfaces easier on the eye. They divide up large open areas and refine edges, corners, and other borders. But you have to be careful. As easily as they as can enhance a chest of drawers, moldings can upset the proportions and throw the whole piece out of visual balance.

More than occasionally, the moldings serve a functional role in furniture construction: hiding joinery or fasteners, masking gaps, providing a physical connection between two separate parts of the construction.

Cutting molding

There are two sorts of moldings: simple and complex. A simple molding is composed of a single basic shape. A complex molding combines two or more of the basic shapes. Here's what this means: Regardless of their size and complexity, all moldings are composed of just a few basic geometric shapes. You can vary the size of the basic shapes and you can vary the way they are arranged. But you won't have much luck trying to invent new ones.

As you devise a molding for your chest of drawers, remember first that a molding is not necessarily a separate strip of wood with a profile cut on it. Quite often, the profile is cut directly on a furniture component—the edge of the top or the base, for example. The basic shapes can usually be produced with a single cutter, and there are cutters available to produce many complex shapes.

On the other hand, a molding is sometimes created from several separate profiled strips that are joined together, as shown in the drawing on p. 22.

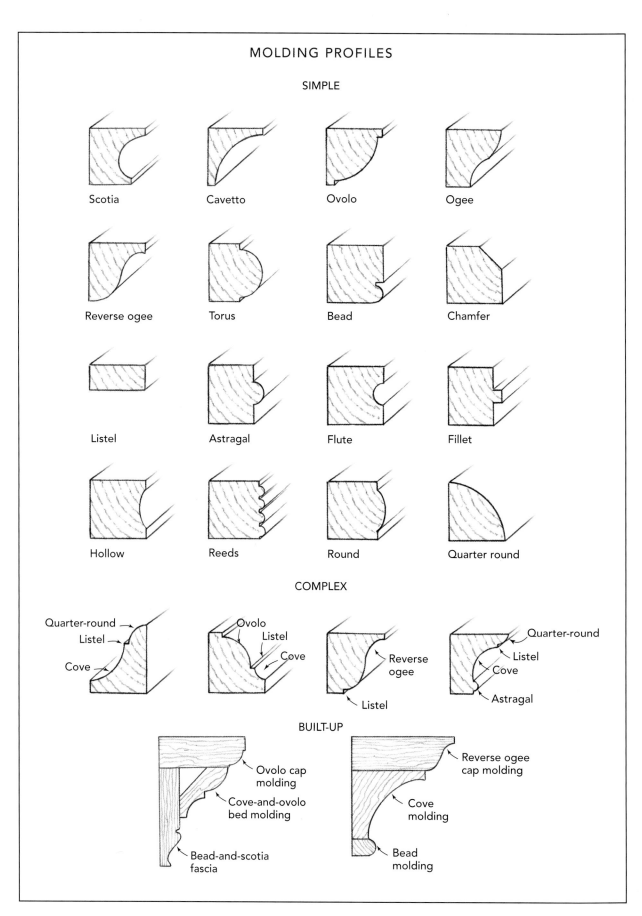

MOLDING PROFILES

SIMPLE

Scotia

Cavetto

Ovolo

Ogee

Reverse ogee

Torus

Bead

Chamfer

Listel

Astragal

Flute

Fillet

Hollow

Reeds

Round

Quarter round

COMPLEX

Quarter-round

Listel

Cove

Ovolo

Listel

Cove

Reverse ogee

Listel

Quarter-round

Listel

Cove

Astragal

BUILT-UP

Ovolo cap molding

Cove-and-ovolo bed molding

Bead-and-scotia fascia

Reverse ogee cap molding

Cove molding

Bead molding

Attaching molding

Good case construction allows sides, top, and bottom to move together. Across the case front, the grain in the molding parallels that in the case, so the molding can simply be glued in place. The sides are different. When you apply a molding to them, it is crossing the grain of the sides. Depending on how it is attached, the molding can restrict the wood's movement, and the resulting tensions can crack the side or pop off the molding.

Over time, cabinetmakers have developed several good ways of attaching moldings cross-grain. In the simplest, the mitered front end is glued and fixed firmly to the case, while the back end is tacked on with wire brads, but not glued. The brads flex as the case side moves, so the molding stays in place. If the molding is a substantial one—like a built-up cornice molding—nails can be used instead of brads. A nail will enlarge its hole at the point where it passes out of the molding and into the case side, allowing movement, while the point of the nail will remain relatively fixed and secure.

Another approach is to mount the molding on a short dovetail keys that are screwed to the side of the case. The back of the molding is slotted from end to end to accommodate the key. It is driven onto the key and then glued to the case and the adjoining molding at the miter. This technique works with large section moldings but not with thin ones.

A variant is to use pan-head screws instead of the dovetail keys. You drive the screws into the side but leave them projecting. Cut a T-slot in the back of the molding to accommodate the screw shank and head. Drive the molding onto the screws and glue it at the miter.

PUTTING IT ALL TOGETHER

Except for building the drawers, which is the subject of the next chapter, I've covered all the basics of building a chest of drawers. All but the aesthetics, that is. Making sure the chest is strong, that the drawers open and close smoothly, and that the piece is scaled to func-tion well for its user are all essential aspects of building a chest. But so is making it a feast for the eyes.

Design is an intimidating realm for many woodworkers. Dressing boards and cutting joints and gluing up assemblies are all logical, concrete processes. But *composing* a chest of drawers is abstract. Working out proportions, dealing with the subtleties of stock thick-nesses, organizing and scaling the drawers, and selecting appropriate moldings and embellishments are all indefinite and mysteri-ous processes. Maybe it is because there are so many options, it is difficult to see the one tree you like in the woods.

Working from existing plans (like those in this book) is a good way to start. You can study the photos and plans and make your own judgments about the success of a design. With a published plan, you can be fairly well assured that the chest is a success from a con-struction standpoint. Beyond slavishly follow-ing the plan, you can copy an existing piece but put your own stamp on it. Change the size or the proportions. Rearrange the drawers or alter the style. Simplifying the design or the construction is valid. You *can* do this as an exercise on paper, but it's better if you actually head into the shop and *build* your version. There's no better way to learn than by doing. You'll see what works and what doesn't.

While you are doing that, you can visit fur-niture stores (take a tape measure along) and museums with good furniture collections. Look at other books and at woodworking and interior-design magazines.

Over time, you'll develop a stockpile of design motifs that appeal to you, that you know work. You'll have a sense, too, of which ones you can use with which other ones. (And which ones don't go together at all.)

The seven chests of drawers in this book display a wide range of forms and design styles. All draw on the past, some more overtly than others. All are good designs; all are well constructed. You may not like every one, but each has something to teach you.

DRAWER-BUILDING BASICS

DRAWERS ARE OBVIOUSLY central to a chest of drawers. Each is essentially an open-topped storage container. Just a box. Function doesn't require a drawer to be fancy or complicated. Typically, we make them of a secondary wood, with just the front made to match (or complement) the chest itself. But we want them to be sturdy and tight, probably not too heavy, and easy to open and close.

If you are a furniture maker, you want a drawer to be something you can construct quickly without sacrificing strength and durability or appearance. This is a bigger challenge than it might appear.

A drawer arguably receives more punishment than any other furniture component. You jerk it open. You slam it shut. Open. *Bang!* Shut. Open. *Bang!* Shut.

A strong, long-lasting drawer needs not only good joinery but also good support in the chest and a way to guide its movement. If it sticks in the case and you need to yank on it to get it to move, you put extra stress on the drawer's joints between the front and the sides. (And you stress the chest itself, too.) Then if you have to throw a hip against it to close it, you are once again stressing the drawer *and* chest.

Traditionally, drawers are constructed and fitted with a lot of handwork. But time is dear, and many a contemporary woodworker favors machine-cut joinery and easy fits. There is, it turns out, no one way to build a drawer.

PARTS OF A DRAWER

Every drawer has the same basic parts: front, back, sides, and bottom. But these parts can be assembled in a variety ways to produce different types of drawers. Curiously, it's not so much the way the drawer is constructed as how the front of it relates to the case that gives the drawer type its name.

The front

The *flush drawer* is easily the most common type. The front of the drawer is recessed within the case so its face is flush with the case facade. To look right, with an even gap all around, the drawer has to *be* right. Moreover, in a chest of drawers, each drawer has to match its neighbors. All need to be flush, all need the same visual clearance around the edges. This makes it the least forgiving type of drawer to the craftsman. The flush drawer is used in the Contemporary Chest (p. 62), the Bow-Front Chest (p. 88), and the Triple Dresser (p. 138).

The *lipped drawer* has a rabbet cut on three edges and sometimes on all four. More often than not, the lip is profiled with a bead. The front nestles partway into the case, and the lip covers the gap between the drawer front and the case. This has the practical benefit of covering up a loose fit. Both the Queen Anne Chest on Frame (p. 166) and the Tall Chest (p. 188) have lipped drawers.

The flush drawer has a front that nestles into the chest, with its face flush with the edges of the chest sides and the drawer dividers.

The lipped drawer has rabbets around the inside face of the front, so it can fit partway into the chest. The lip overlaps the sides and dividers, giving the chest facade depth.

The overlay drawer is a modern contrivance, calculated to expedite production and facilitate the use of manufactured runners. A "show" front attached to the drawer box overhangs the box to conceal the runners and to overlay the edges of the case.

The third type is the *overlay drawer,* in which the front overlays the edges of the case, concealing it. Often—but not always—the front is an element distinct from the drawer box, one attached after the box is assembled. It may be attached with adjusters, which are eccentrics that allow the front to be shifted up and down, side to side, or even cocked slightly. With this style, the drawer builder can deal with alignment by adjusting just the front and not the entire box and its support system. The Double Dresser (p. 116) has this style of drawer.

Regardless of type, the drawer front is invariably made of the primary wood used in the chest. In any of the constructions, it can be an integral part of the drawer box or added on as a false front.

The sides and back

The drawer sides do a lot of work. Together with the front and back, the sides form the walls of the drawer box. Usually, the sides provide the main support for the bottom, either directly or through slips, which are glued to them.

In traditional drawer systems, the sides support the whole drawer as well, since its weight is transferred through the bottom edges to the runners that are built into the chest. And there's more. The drawer's movement depends on the sides. Those bottom edges are the bearing surfaces on which the drawer moves. The outer faces of the sides are the guides that rub against the chest walls, keeping the drawer on a straight course—you hope.

The back, in contrast, does little beyond linking the sides and enclosing the drawer box. In the typical traditional drawer, the bottom is secured to the back to keep it in place and to keep it from sagging in the middle.

But when the drawer moves, the back is just a passenger. Not only is it set on top of the drawer bottom so it doesn't drag on the runners but, in some designs, it is deliberately held below the top edges of the sides so it doesn't drag on the runners overhead either.

The bottom

As with drawer fronts, there are several types of drawer-bottom construction, which you use depends on the material, the style of the drawer and chest, and the size of the drawer and the strength needed.

Most common is the *open-back construction,* in which the bottom is fitted beneath the drawer back into grooves in the drawer sides and front. This construction is almost essential if you plan to use a solid-wood bottom.

The advent of effectively stable materials— plywood and hardboard, specifically—made the *fully enclosed construction* reasonable. Here, the bottom is housed in grooves in the back as well as in the sides and front.

A primitive, seldom seen type is the *overlay construction.* Here, the bottom is a panel that's laid over the edges of the sides, front, and back and nailed into place. The durability and wearability of this construction are questionable (although some very old pieces—now in museums—have drawers built this way).

An interesting, yet uncommon hybrid is the *NK construction.* This bottom is composed of two shoe plates and a bottom panel. It is then assembled and fitted to the chest; finally, the drawer box is glued to it. This unfamiliar bottom construction is used on the Bow-Front Chest (p. 88).

The pull

Don't overlook the part that's used to open the drawer. It is a handle or a knob or a finger grip of some sort. It can be wood, metal, or plastic, purchased or shopmade. Collectively, these parts are called *pulls.* Seldom can you get away without at least one pull on each drawer; wide drawers usually have two.

WOOD CHOICES

Woodworkers are pretty acclimated to the idea of making drawers from an assortment of materials.

The front is the primary wood, of course. But rarely are the sides and back made from the primary wood. We *know* we can save a little money by using a less-spectacular, less-costly wood for the drawer sides and backs. We use this secondary wood for all the non-public parts in a chest. In some contemporary chests, the drawer sides and backs (and structural fronts) are cut from plywood.

Then there's the bottom. It's traditional to make drawer bottoms from thin pieces of the secondary stock. But these days, plywood is used for the drawer bottoms more often than not. It is inexpensive, strong, and lightweight. Plus you can transform a sheet of plywood into a stack of drawer bottoms in a matter of minutes.

What are the criteria for selecting materials for non-show parts of the drawers?

Strength and weight

The trick is finding the balance between weight and strength. Maple drawer sides are very strong, of course, but they will add considerably to the weight of the dresser. Pine drawer sides will make for a lighter case, but they will need to be somewhat thick. The thick sides may be an advantage, depending on the method of support. Side mounts, a system that requires grooves to be cut in the drawer sides for runners that are attached to

Without a knob or handle, a drawer can be near impossible to open. These pulls—some handcrafted, most manufactured—barely suggest the range of options available.

the case, demand that the drawer sides be on the thick side. This also means the drawer will be fairly heavy. A drawer that rests on top of the runner, on the other hand, can be slimmed down.

Aesthetics enters this picture too. You may not be fully conscious of it, but you do have notions about appropriate proportions for drawers. One that's too bulky or too slender for its dimensions does get your attention. A smaller drawer—one that's one-half or one-third the width of the case—is proportioned accordingly. The sides and back are thinner and so, sometimes, is the bottom. The thickness of these parts are bulked up proportionally in a deep, full-width drawer housed in the same case.

Stability and wearability

The traditional drawer opening—the one bounded by the case sides and the drawer dividers above and below—is, in effect, a frame that doesn't vary appreciably in dimension from season to season. A drawer made of solid wood *does* change appreciably in dimen-

sion from season to season. And if too little overhead clearance is allowed, the drawer will stick in humid summers.

The upshot: Select your secondary wood based on its stability and the way it is sawed. Certain woods move more and are more prone to various types of distortion with seasonal humidity changes than others. These you want to avoid, so you eschew the woods that move the most. In addition, quartersawn lumber is much more dimensionally stable than flatsawn lumber. If possible, take your drawer sides and backs from quartersawn stock.

Plywood, of course, isn't prone to any of these woes. It is stable, strong, and lightweight. But it conflicts with most people's aesthetic sense. The show of plies at the edges doesn't suit. If the aesthetics isn't a problem and if speedy production is a goal, plywood is great. The time spent making the sides and back is reduced—no jointing, no resawing, no planing. You won't want to dovetail the parts together; but speedy production and hand-cut dovetails are on different pages anyway. You'll use a machine-cut joint.

Wearability is a different measure. In a chest of traditional construction, the drawer rests on a frame composed of the drawer divider, runners, and perhaps a back rail. The bottom edges of the sides are the bearing surfaces. If you use a soft wood for the sides and/or the runners mounted in the chest, the drawer will wear quickly. The edges of the sides deteriorate, and grooves may be worn in the runners (and even into the drawer divider).

The goals here are to use a reasonably durable wood as your secondary and to use the same species for both the drawer sides and the runners. Good choices include poplar, soft maple, and alder.

It's worth mentioning too that in addition to wearing faster, soft woods slide more sluggishly. However, traditional cabinetmakers in the United States frequently used softwood drawer sides because of the ready availability.

Cost and labor

Cost is the main rationale behind the use of a secondary wood. The poplar used in the chests I built for the photos cost only 20 percent or 25 percent of what I spent for the cherry, walnut, and hard maple.

But the material expense is only part of your cost calculation. Here I'm thinking primarily about drawer bottoms. I pointed out that in just minutes, you can produce a stack of drawer bottoms from a sheet of plywood. How long will it take to make a matching stack of solid-wood drawer bottoms?

That job usually entails resawing as well as the usual labor for prepping the materials. Glue-ups typically are necessary to get panels 15 in. to 18 in. wide, which drawers in a large chest require. And joinery cuts will be required in the bottom itself, either a rabbet or some sort of panel-raising operation.

DRAWER CONSTRUCTION

Everybody likes to open drawers and see what's inside. But woodworkers usually look at the joints first and will probably slide the drawer in and out a couple of times to gauge its fit in the case and the smoothness of its action. And *then* they'll look at the contents.

Such assessments reveal the aspects you need to keep in mind as you select the joints and constructions you'll use in building drawers for a particular chest. Looks are important.

All sorts of joints are used in drawers, from the traditional dovetail to the nailed butt. In my mind, the strongest joint needs to be between the front and sides. This is where the stress hits, every time a drawer is opened or closed. This is also the joint that needs to look good, because it is the one that's seen each time the drawer is opened.

The joint between the sides and back needs to be strong too, of course. But most of the stresses on it are secondhand, more inertial than direct. It's seldom seen since you have to completely remove the drawer from the case to look at it. Function is more important than looks here.

Front-to-side joinery

The front-to-side joints take the bulk of the strain on a drawer. If you try to open a badly built drawer, you may come away with just the drawer front in your hand (see the drawing on p. 30).

Dovetails generally indicate a well-made drawer. The half-blind dovetail is *the* traditional joint for this application. It's has been the joint of choice for literally centuries. Two hundred years ago the hand-cut dovetail was just about the only joinery option for drawers. It was used on low-end furniture as well as high. Now that there are many other machine-cut joint options, half-blind dovetails are the seen primarily on high-end and custom-made drawers.

The half-blind dovetail doesn't show to the front, but when the drawer is opened, it makes a great impression. It can be used for any of the three types of drawers (lipped, flush, and overlay—remember?), though a false front is necessary for an overlay drawer.

If the half-blind dovetail has drawbacks, they stem from the effort it takes to make them. They are time-consuming to cut by hand and finicky to fit. You can use a router

FRONT-TO-SIDE JOINERY

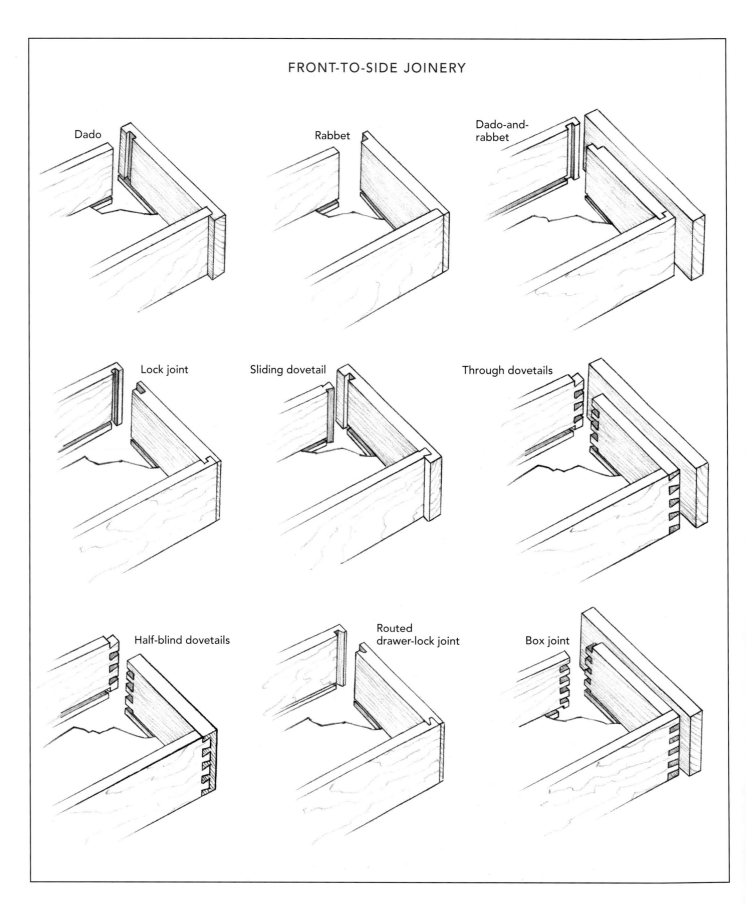

Dado

Rabbet

Dado-and-rabbet

Lock joint

Sliding dovetail

Through dovetails

Half-blind dovetails

Routed drawer-lock joint

Box joint

and one of several jigs to machine them, but dialing in the proper settings for jig and router can take time. And even with the most adjustable of the jigs, the results are pretty clearly machine cut.

The other dovetail variants are not universally acceptable for front-to-side joinery. Through dovetails are strong, but they show to the front as well as the side. If exposed joinery is part of the design, then that's okay. Other–wise, a false front is needed to conceal them.

The sliding dovetail is strong and easy to make (once you have the setup), but you can't cut the dovetail slot close by an edge. Thus it will work only on an overlay drawer (or a flush drawer riding on commercial side-mounted slides). You can produce through and stopped joints; in the former, the dovetail slot is visible in the top edge of the drawer front, in the latter it is not.

A joint that *looks* akin to a dovetail is the box joint. It's strictly a machine-made joint (cut using a router or table saw), and it doesn't have as sophisticated an interlock as the dovetail. The many gluing surfaces make up for that and yield a joint that's plenty strong for drawers. You can even make a half-blind box joint, though it isn't used on any of the chests in this book. You can use the box joint in the same functional (but not aesthetic) applications you would the dovetail.

At the opposite end of the drawer joinery spectrum you'll find the rabbet and the dado joints. The advantage of a plain rabbet or dado for joining a drawer front to the sides is ease of construction. Neither joint has any interlock that's integral to the joint, and there's no good gluing surfaces, so you shouldn't expect the drawer to survive for generations.

A hybrid, the dado-and-rabbet joint, does lock the parts together and is easy to make. But it exposes the end-grain of the drawer side to the front of the drawer, so it needs a false front for all but the most utilitarian applications.

Where construction efficiency is paramount, the lock joints are worth serious considera-tion, especially the routed drawer lock. These joints are strong and simple. The routed joint is cut with a special bit, which produces both parts of the joint. The lock joints work equally well on overlay and flush drawers and can be used to produce lipped drawers as well.

Side-to-back joinery

As noted previously, function is more significant in the side-to-back joinery than appearance. It is quite common to find one joint used at the front, and a very different one at the back (see the drawing on p. 32).

Historically, through dovetails were used at the back of a drawer. In custom work, they are still the joint of choice. But it's common these days to join the backs and sides with less fuss—a dado, dado-and-rabbet, even a nailed butt joint may be suitable.

If you're making the front joints with a particular machine setup—a routed lock joint, for example—it's practical to make the back joints the same way.

Bottom construction

The bottom keeps the drawer's contents from falling on through. So the bottom itself has to be strong enough to support whatever you put in the drawer. The joinery between the bottom and the walls of the drawer also must be strong (see the drawing on p. 33).

The first issue to settle when building the bottom is the joinery. Almost without exception, drawer bottoms are housed in grooves cut in the drawer's front and sides—and sometimes in the back as well. Just bear in mind that the groove compromises the strength of the side at the most critical location. A groove that's too wide or too deep carries—along with the bottom—the potential for failing. And a thin side simply sharpens the dilemma.

A traditional solution to the problem is the drawer slip. Drawer slips are basically square strips of wood glued to the sides at their bottom edges. The grooves for the bottom are cut in the slips. A reasonably sized groove isn't going to compromise the material.

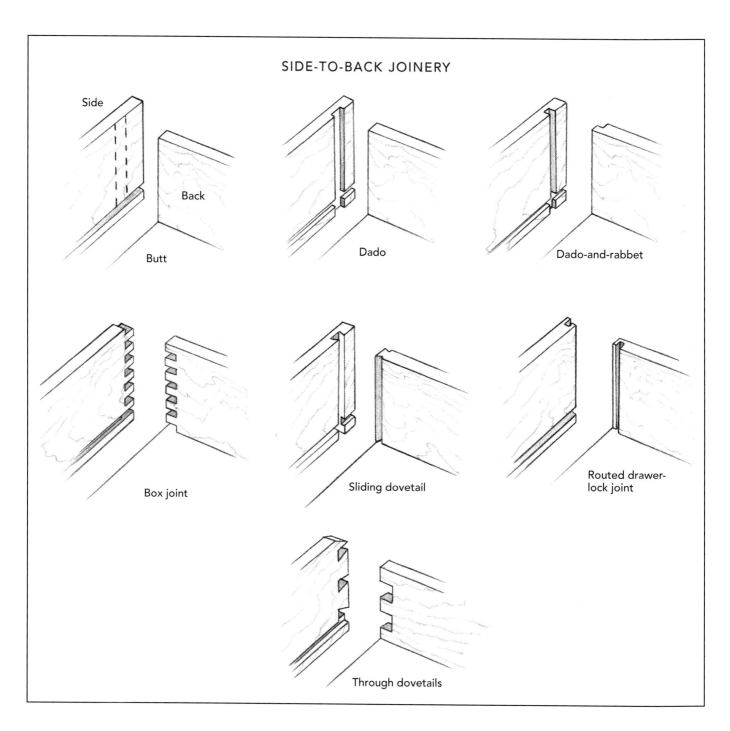

SIDE-TO-BACK JOINERY

Side

Back

Butt

Dado

Dado-and-rabbet

Box joint

Sliding dovetail

Routed drawer-lock joint

Through dovetails

Slips have an additional benefit. Thin sides that slide on runners gradually wear down over the years, detracting from a good fit. Drawer slips increase the bearing surface and thereby extend the useful life of the drawer.

Once you've settled on a joinery technique, you'll need to decide between plywood and solid wood for the bottom. Plywood tends to be the choice for all but the traditional chest of drawers. At any given thickness, it is stronger than solid wood. It is stable, so movement isn't a problem. In fact, it can be glued in place, which helps stiffen the box. And the economics of plywood are excellent.

The primary drawback of plywood is the actual thickness. A ¼-in. sheet is really about 7/32 in., and even that is an average across the board. If you cut a ¼-in. groove for the stuff,

DRAWER BOTTOM CONSTRUCTION

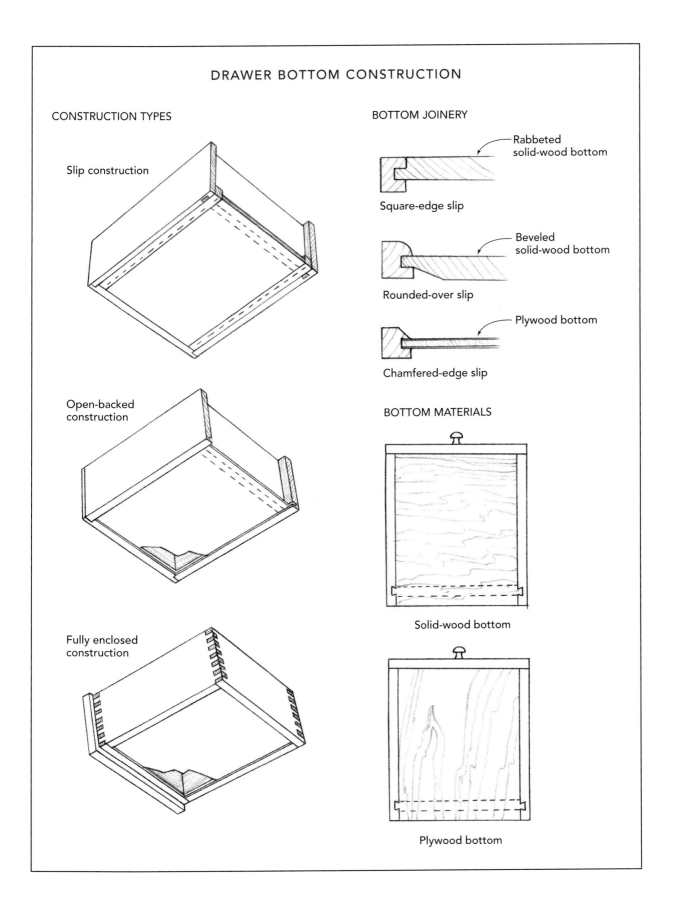

CONSTRUCTION TYPES

Slip construction

Open-backed construction

Fully enclosed construction

BOTTOM JOINERY

Rabbeted solid-wood bottom

Square-edge slip

Beveled solid-wood bottom

Rounded-over slip

Plywood bottom

Chamfered-edge slip

BOTTOM MATERIALS

Solid-wood bottom

Plywood bottom

Two small drawer-bottom panels are stronger than one large one. A muntin divides the drawer box's bottom (just the way it divides a window) so two panels can be used instead of one.

it'll rattle. The solution usually is to use a cutter that's less than ¼ in. and to make two passes to match the groove width to the sheet thickness.

The alternative is the traditional solid-wood bottom. Unless the bottom is very small, it needs to be thicker than ¼ in. (thin wood is prone to crack). A typical solid-wood bottom thickness is ½ in., though smaller drawers might have ⅜-in. bottoms. Some furniture makers favor ⅜-in. bottoms.

To reduce the width of the groove required, the solid-wood bottom needs a tongue or a rabbet. Use a panel-raising bit in a table-mounted router to mill the bottoms, and you'll get a nicely formed tongue to fit the grooves.

A solid bottom should be used only on open-back construction, so the bottom can expand and contract. Orient the bottom so its grain runs side to side, parallel to the back. To ensure that the bottom can expand and contract, use a screw (or a nail) in a slotted hole when securing the back edge of the bottom.

On a very wide drawer a large, one-piece bottom is likely to sag, and it may eventually break. You can deal with this before it becomes a problem by adding a center muntin. This frame piece, which extends from front to back, divides the bottom opening of the drawer box so two smaller panels can be used to form the bottom. The muntin must be grooved like the sides, and it must be securely anchored to the front and back. You can use a tongue or dovetail at the front. At the back, cut a rabbet across the muntin to form a simple lap joint between it and the drawer back.

FINISHING THE DRAWERS

Finishing is a topic largely left unexplored in this book. But a few words on finishing drawers are essential.

Drawers often are left unfinished. Oil-based finishes, which are favorites of furniture makers, give the insides of drawers a vaguely rancid smell. That's because the drawer boxes are usually closed, inside the chest, where air circulation is negligible. This odor can persist for years and years. It can permeate clothes kept in the drawers. Some people don't notice it, others are extremely conscious of it.

Being wary of this problem, some furniture makers avoid using *any* varnish (not only oil-based ones) on the body of a drawer and even on the inside of a chest of drawers.

If you want to finish the drawers in your chest, try shellac. A couple of diluted coats will dry quickly and seal the wood. Then sand lightly with 220-grit paper to eliminate the nibs; finally, apply paste wax.

SUPPORTING THE DRAWERS

A drawer alone—just an open box—is an oddity. For it to work as intended, it has to be installed in a case in a way that allows it to be opened and closed. The movement must be smooth; and once open, the drawer has to be able to stay open without your help.

Drawer movement can be controlled in several ways. Some mounting systems are integral to the case, and others are add-ons. Regardless, the mounting system should be carefully planned along with the case and drawer design.

Runners, guides, and kickers

The most common approach to supporting a drawer is an arrangement of a drawer divider and runners (see the drawing on p. 36). The divider is a rail extending from side to side. It separates the drawers visually and physically. And it also supports the front end of the runners.

Attaching the runners directly to the case sides seems simple. But, of course, runners can't be glued to solid-wood sides, because they'll restrict the sides from expanding and contracting. Instead, the runners are set into dadoes and glued at one end only. Or they are attached with screws in slots. Or they're housed (unglued) in dovetail or dado slots. A long-standing practice is to capture the (slightly short) runners (unglued) between the front divider and back rail, both of which are glued in place.

A frame-and-panel chest or a case with a face frame requires an additional element—the drawer guide—to limit side-to-side movement of the drawer.

Side-by-side drawers, often included in dressers and other chests, need support in the middle of the case, away from the sides. The usual approach here is to suspend a wide runner between the drawer divider in front and a rail in back. A vertical divider with a guide behind it separates the neighboring drawers.

An important element in most drawer-mounting systems is the kicker. A kicker prevents the drawer from tipping down as it is opened. It is just like a runner but, generally, is mounted above the drawer side. A single center kicker may be used for a top drawer.

Side mounts

Some furniture designs make it difficult to use runners. A case that has no dividers separating the drawers is an example. In this situation, you can use side-mounted slides. The slide is a strip of wood attached to the case side. Grooves for the slides must be cut in the drawer sides.

All the caveats about mounting a runner to a solid-wood case side apply here. This can be a drawback to the system. Another significant drawback is that the drawer sides need to be quite thick to be able to accommodate the slide.

Center runners

Wide drawers supported by side runners have a tendency to cock slightly as they are moved and to bind. The wider the drawer, the more likely it is that this will happen.

A single center-mounted runner and guide is the solution. The runner, which is attached to the underside of the drawer, has a channel in it that rides over a guide that's attached to the apron or web frame (see the drawing on p. 36).

Runners and Guides

SOLID WOOD/PLYWOOD CASE CONSTRUCTION

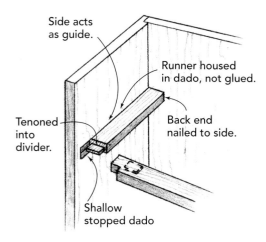

Side acts as guide.

Runner housed in dado, not glued.

Tenoned into divider.

Back end nailed to side.

Shallow stopped dado

The runner is housed, unglued, in a dado cut into the chest side. The chest side serves as the drawer guide.

POST-AND-PANEL CONSTRUCTION

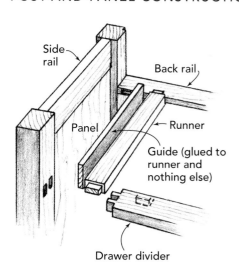

Side rail

Back rail

Panel

Runner

Guide (glued to runner and nothing else)

Drawer divider

The runner tenons fit into mortises in the drawer divider and back rail. The drawer guide is glued to only the runner.

CENTER RUNNER AND GUIDE

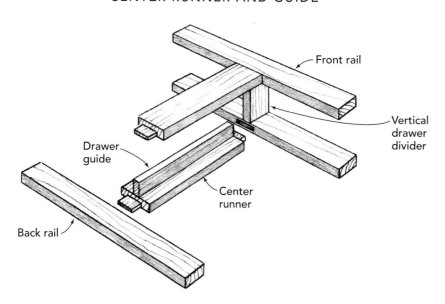

Front rail

Vertical drawer divider

Drawer guide

Center runner

Back rail

The runner tenons fit into mortises in the front and back rails. The drawer guide is glued to the runner.

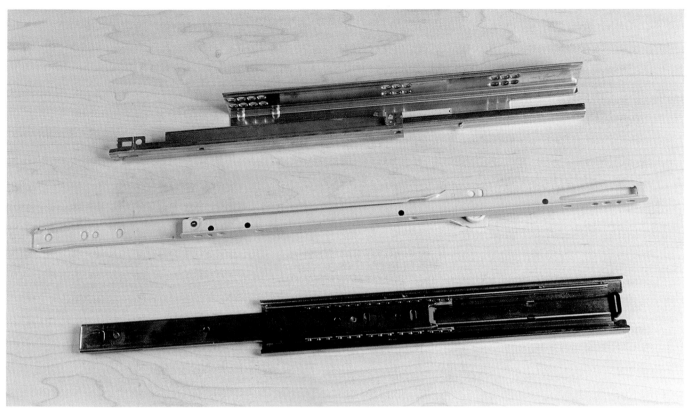

Manufactured runners

Metal slides with ball-bearing wheels are another drawer-mounting technique. The slides are mounted in pairs to the case and drawer or singly under the center of the drawer. They offer a smooth opening-and-closing action that's not affected by wood movement. They can be used in chests of drawers, just as they can in most furniture applications. Full-extension slides allow the full depth of the drawer to be exposed, something you can't get with the other drawer-mounting techniques.

Drawer stops

Drawer stops keep all styles of drawers from falling out of their cases (opening stops) and flush drawers from sliding too far into their cases (closing stops).

A turn button is the simplest opening stop. It can be mounted on the inside of the drawer back or on the back edge of the front rail. Pivoting it out of the way allows the drawer to be inserted or removed.

A small block of wood tacked or glued to the back of the runner is the easiest way to make a closing stop. With the back removed and each drawer inserted so it's perfectly flush with the cabinet face, apply the closing stops with a dab of glue. Then add a couple brads or a small screw. You can also mount the closing stops onto the front rail, so they will catch against the back of the drawer front. They're definitely harder to locate and attach here, but such a stop can work for both opening and closing.

Manufactured drawer runners range from sturdy side mounts to discrete concealed runners that hide beneath the drawer box. Low-cost runners mount to the bottom edges of the drawer sides.

PENNSYLVANIA DUTCH CHEST

This chest represents the origin of the chest of drawers. Its antecedent is the basic chest—and with the addition of more drawers and numerous details it eventually evolved into the varied form we know today. This reproduction, made Harry L. Smith of Kunkletown, Pennsylvania, is typical of the form and has two side-by-side drawers as well as an interior till with a lid for holding small items.

Rural New Englanders and the Shakers also made these hybrid chests. Initially, they used a single deep drawer in the piece. Later examples have two or three drawers of graduated depth, in a stack. These later examples link clearly to the traditional chest of drawers.

As a woodworking project, this chest is rewarding. Beyond the dovetail joinery, it is fairly simple to construct. And it does have drawers. The outcome is a practical piece of furniture.

If you use an attractive hardwood, it will look great under a clear finish. And the paint? It's true to its antecedents in form and colors. It looks new because it is. Allowing the finish to be bright and fresh, rather than distressing it to look old, is Smith's way. "With time, it will develop its own story," he says.

Pennsylvania Dutch Chest

THE CHEST PART of this piece fits inside the base. Screws driven through the rail and cleats join the two sections.

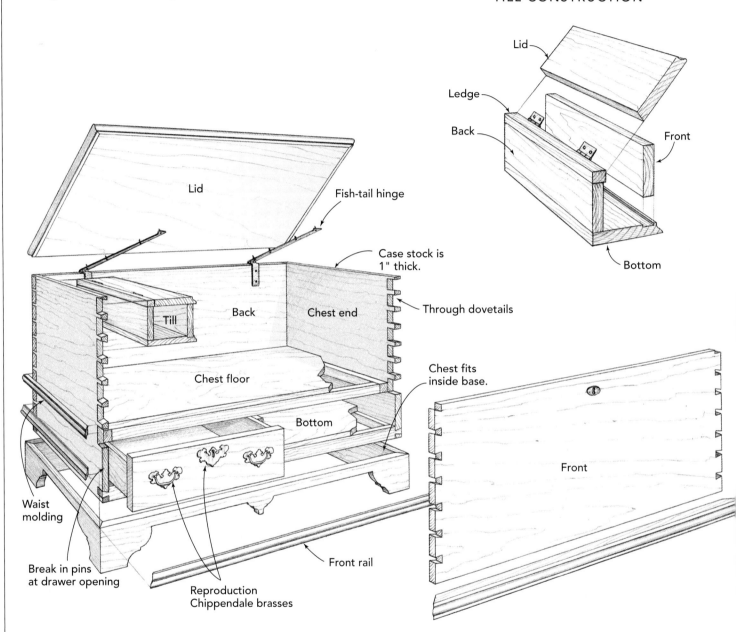

Lid

Ledge

Back

Front

Bottom

Lid

Fish-tail hinge

Case stock is 1" thick.

Back

Chest end

Through dovetails

Till

Chest floor

Chest fits inside base.

Bottom

Front

Waist molding

Break in pins at drawer opening

Reproduction Chippendale brasses

Front rail

Front

SECTION VIEW

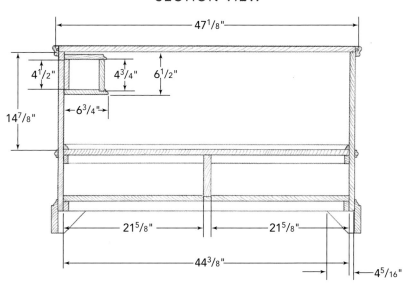

END VIEW

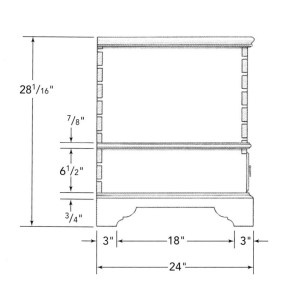

FRONT VIEW

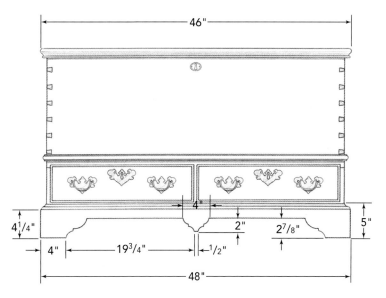

BUILDING THE CHEST STEP-BY-STEP

CUT LIST FOR PENNSYLVANIA DUTCH CHEST

Chest

2	Ends	$^{13}/_{16}$ in. x 23$^{15}/_{16}$ in. x 23 in.	clear white pine
1	Front	$^{13}/_{16}$ in. x 16$^{13}/_{16}$ in. x 46 in.	clear white pine
1	Back	$^{13}/_{16}$ in. x 23$^{15}/_{16}$ in. x 46 in.	clear white pine
1	Box floor	$^{13}/_{16}$ in. x 21$^{5}/_{8}$ in. x 44$^{3}/_{8}$ in.	clear white pine
1	Bottom	$^{13}/_{16}$ in. x 22$^{3}/_{16}$ in. x 44$^{3}/_{8}$ in.	clear white pine
2	Cleats	$^{13}/_{16}$ in. x 1$^{1}/_{4}$ in. x 21$^{5}/_{8}$ in.	clear white pine
2	Cleats	$^{13}/_{16}$ in. x 1$^{1}/_{4}$ in. x 42$^{3}/_{4}$ in.	clear white pine
2	Cleats	$^{13}/_{16}$ in. x 1$^{1}/_{2}$ in. x 21$^{5}/_{8}$ in.	clear white pine
1	Cleat	$^{13}/_{16}$ in. x 1$^{1}/_{2}$ in. x 42$^{3}/_{4}$ in.	clear white pine
1	Drawer partition	1$^{1}/_{8}$ in. x 6$^{5}/_{16}$ in. x 22$^{3}/_{16}$ in.	clear white pine
1	Front rail	$^{13}/_{16}$ in. x 1$^{1}/_{2}$ in. x 46 in.	clear white pine
2	Quarter-round moldings	$^{3}/_{4}$ in. x $^{3}/_{4}$ in. x 21$^{5}/_{8}$ in.	clear white pine
2	Quarter-round moldings	$^{3}/_{4}$ in. x $^{3}/_{4}$ in. x 44$^{3}/_{8}$ in.	clear white pine
1	Till bottom	$^{3}/_{4}$ in. x 6$^{3}/_{4}$ in. x 21$^{5}/_{8}$ in.	clear white pine
1	Till front	$^{3}/_{4}$ in. x 4$^{3}/_{4}$ in. x 21$^{5}/_{8}$ in.	clear white pine
1	Till back	$^{3}/_{4}$ in. x 4$^{1}/_{2}$ in. x 21$^{5}/_{8}$ in.	clear white pine
1	Till hinge ledge	$^{3}/_{4}$ in. x 1 in. x 21$^{5}/_{8}$ in.	clear white pine
1	Till lid	$^{3}/_{4}$ in. x 5$^{7}/_{16}$ in. x 21$^{1}/_{4}$ in.	clear white pine

Base

1	Base front	1 in. x 4$^{1}/_{4}$ in. x 48 in.	clear white pine
2	Base ends	1 in x 4$^{1}/_{4}$ in. x 24 in.	clear white pine
2	Corner blocks	2 in. x 2 in. x 3$^{5}/_{16}$ in.	clear white pine
2	Rear brackets	$^{13}/_{16}$ in. x 3$^{5}/_{16}$ in. x 4$^{5}/_{16}$ in.	clear white pine
2	Corner blocks	1 in. x 1 in. x 3$^{5}/_{16}$ in.	clear white pine
1	Drop blocking	$^{13}/_{16}$ in. x $^{13}/_{16}$ in. x 3 in.	clear white pine
1	Front base molding	$^{7}/_{8}$ in. x $^{13}/_{16}$ in. x 48 in.	clear white pine
2	End base moldings	$^{7}/_{8}$ in. x $^{13}/_{16}$ in. x 24 in.	clear white pine
1	Front waist molding	$^{5}/_{8}$ in. x $^{7}/_{8}$ in. x 47$^{1}/_{4}$ in.	clear white pine
2	End waist moldings	$^{5}/_{8}$ in. x $^{7}/_{8}$ in. x 23$^{5}/_{8}$ in.	clear white pine

CUT LIST FOR PENNSYLVANIA DUTCH CHEST

Lid

1	Lid	1³/₁₆ in. x 23⁵/₁₆ in. x 47⅛ in.	clear white pine
1	Front lid molding	1⁵/₁₆ in. x 1½ in. x 48 in.	clear white pine
2	End lid moldings	1⁵/₁₆ in. x 1½ in. x 24 in.	clear white pine

Drawers

2	Drawer fronts	1³/₁₆ in. x 5⁵/₁₆ in. x 22³/₁₆ in.	clear white pine
4	Drawer sides	⅝ in. x 4¹¹/₁₆ in. x 20¹⁵/₁₆ in.	clear white pine
2	Drawer backs	⅝ in. x 4 in. x 21⅞/₁₆ in.	clear white pine
2	Drawer bottoms	⅝ in. x 20¾ in. x 20¹/₁₆ in.	clear white pine

Hardware

1 pair	Butt hinges	1½ in. x 1½ in.
1 pair	Fish-tail hinges	from Fisher Forge or from Horton Brasses
1	Half-mortise chest lock with brass case	from Horton Brasses; item #CL-2P
1	Rosette horizontal escutcheon	from Horton Brasses; item #RTE-3
2	Half-mortise drawer locks	from Horton Brasses; item #LK-2
2 pairs	Chippendale brass drawer pulls with 2½-in. borings	from Horton Brasses; item #H-34S
2	Chippendale brass escutcheons	from Horton Brasses; item #H-34SE

THE CHEST HAS FOUR major components: the chest box itself, the base, the lid, and the drawers. The chest, obviously, is the first component to build. Make the base next, followed by the lid and finally the drawers.

BUILDING THE BOX

Gluing up the wide panels and cutting the dovetails are the most demanding processes in the chest's construction. But the work begins with stock preparation. Smith built his entire chest of clear white pine. Especially if you plan to paint your chest, pine and poplar are good lumber choices.

Preparing the stock

Preparing the stock isn't the most rewarding work, but it is essential. The stock you use must be flat and true, milled to a consistent thickness. If you have an out-of-the-way storage spot, you can dress all the stock for the project at once. Set aside the parts you aren't using immediately—like the lid and drawer and base parts—and sticker them.

1. Lay out the parts on the stock. Use rough dimensions at this stage, marking off pieces that are 2 in. or 3 in. longer than the cut list dimensions and somewhat wider. Work around knots and other defects. If you cut stock for the moldings now, be sure to cut the pieces 2 in. to 3 in. wide, so that you can later mill the profile on both long edges.

2. Cut the parts. This is primarily a cross-cutting stage. The very narrow pieces, like the cleats and the front rail can be ripped from wider boards at a later stage. The goal here is to reduce the length of boards as much as possible before jointing and planing.

3. Joint one face of each piece. If the rough-cut boards are cupped, it's easiest to joint the concave face. Simply get it flat. There's no need to mill the face smooth and free of saw marks.

4. Joint an edge of each piece. You need one straight edge, so pick the edge that will be easiest to straighten. If the board is crooked, it usually is easiest to deal with the concave edge. Make sure the jointer's fence is square to the tables. Hold the flattened face of the stock firmly against the fence and feed it over the knives.

5. Now plane and thickness each board. There are two goals here. The first is to mill the still-rough face flat and parallel to the jointed face. The second is to reduce the overall thickness of the board to some predetermined dimension. Most of the stock used in the chest is $^{13}/_{16}$ in. thick.

6. Rip the boards to width. Do this on the table saw, of course. And reference the rip fence with the jointed edge.

Gluing up the panels

The box has a lot of wide panels—panels that were single boards back when the original was built. Unless you have a remarkable cache of lumber, you'll have to edge-glue narrow boards to form the wide ones required. Glue up the boards for the box walls first; then do the boards for the box floor and box bottom. While you're at it, glue up the boards for the large chest lid, too.

Edge-gluing stock is one of woodworking's most basic procedures; and everyone has favorite tricks that work for him, such as incorporating splines or biscuits in the joints to expedite alignment. If you are relatively new to the task, here is a basic routine:

1. Select the boards to make up the panel. Lay them out on the workbench and decide how you'll arrange them for gluing. When you've settled on an arrangement, draw a large triangle across the face of the panel-to-be. The triangle will enable you to put the boards back in this same arrangement when you glue them together.

2. Joint the edges, if you haven't already.

3. Set out bar or pipe clamps. For long panels—like the 46-in.-front—use a minimum of five clamps. Set three out on the benchtop: at the

Photo A: When gluing narrow boards edge to edge to form a wide panel, the first clamp to tighten is the center one. Alternate the clamps, so they span both faces of the panel to help keep it flat.

center and at each end. The other two clamps are applied across the top of the panel, on either side of center. Adjust the jaw positions to accommodate the panel and a caul strip along each edge.

4. Apply the glue. Lay the boards, triangle up, across the bars of the clamps and tip them up on edge. Squeeze a bead of glue onto the "up" edge of each board, running it from end to end. Then paint the edge, spreading the glue in a nice thin coat over the entire edge with a small brush.

5. Begin tightening the clamps. Start with the center one. As you tighten this clamp, work the boards into alignment, so the top surface is flush all the way across. This may require lifting or pushing down on the ends of the individual boards to get the surfaces lined up just at the center.

6. Apply the intermediate clamps across the top surface (see **photo A**). Tighten them; and as you do, keep the boards flush (or work the ends up or down to make them flush).

7. Tighten the end clamps last. If the very ends of the board are still slightly up and down, leave the bar clamp slack and apply a metal clamp or a handscrew directly on the offending joint to force the boards into line; then tighten the bar clamp and remove the clamp from the end of the panel.

8. Wash off glue squeeze-out. A good spread should produce a fairly even line of glue beads. Runs indicate that you applied too much glue, and these you especially want to wash off before they skin over. Use a wet rag for this—not just damp, but wet.

9. Set the glued-up panel aside while the glue cures. Yellow glue, which may be the adhesive most used in the woodworking shop, requires only ½ hour of clamping time but a full 24 hours of curing time. To expedite the project, you can pop the clamps after ½ hour to 1 hour, freeing them for use on the next panel. But set the panel aside overnight to allow the glue to cure completely.

Cutting the dovetails

The box walls are joined with through dovetails, as shown in "Box Construction" on p. 46. Those that join the back and ends are straightforward. Between the front and the front rail and the ends, there is a minor

Tip: A complete lack of glue squeeze-out means you may not have applied enough glue, leaving a glue-starved joint that is inherently weak. Rather than hope for the best, see if you can remove the clamps, separate the boards, and redo the glue-up.

BOX CONSTRUCTION

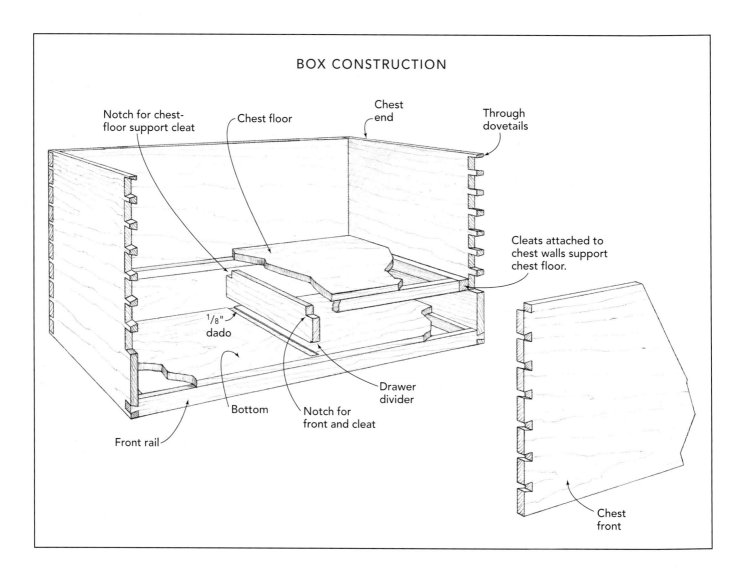

Notch for chest-floor support cleat

Chest floor

Chest end

Through dovetails

Cleats attached to chest walls support chest floor.

$\frac{1}{8}$" dado

Drawer divider

Bottom

Notch for front and cleat

Front rail

Chest front

complication: the undovetailed section of the ends adjacent to the drawers. When cutting the pins on the front end of the end panels, in other words, you work a section matching the width of the front panel and you cut a pair of half-pins at the bottom for the front rail. And the 5⅝-in.-wide section between them is left uncut.

How you size, lay out, and cut the dovetails is up to you. Many woodworkers lay them out and cut them by hand with dovetail saw, chisels, and a mallet. Smith routs his with the aid of a Leigh jig. I followed Smith's lead when making my chest (see **photos B** and **C**).

If you will be routing these through dovetails with the Leigh jig, here are some tips.

1. Lay out both pin boards (which are the box ends). Make the layout as symmetrical as possible. Use the pin boards to set the jig's guide fingers.
2. Mark the joints, labeling the outside faces and the mating parts.
3. Rout the mates to each joint in turn. That is, rout the pins for the left rear of the box, followed by the tails for that joint. Then do the pins for the right rear joint, followed immediately by its tails. And so on.
4. Remember to rout the tails on the front rail when you do the tails on the box front.

Photo B: Most router dovetail jigs require you to cut both pins and tails the with workpiece clamped vertically. Thus to cut the tails on the nearly 4-ft.-long chest front means the jig must be elevated well above benchtop height.

Photo C: The two ends are mirror images, not duplicates. Cut the pins on the front edge of the first chest end. Then immediately cut the tails on the correct end of the front and front rail.

5. After doing the first front joint, you must completely rearrange the jig's guide fingers to do the second front joint. They are mirror images, not duplicates.

Take your time. Check your setups and parts orientations two or three times because you can cut only once.

Assembling the box

With the dovetail joints cut, move ahead with assembly of these parts. The first step is a complete dry assembly.

1. Dry-assemble the box. Stand the back on its top edge, perhaps applying a handscrew to one end as a foot. Slide an end into position, line up the pins in the slots, and gently tap the joint closed. Remove the handscrew and join the second end to the back. Stand the front on its top edge, move it into position, line up the joints, and tap it into place (see **photo D** on p. 48). Finally, join the front rail to the assembly.

2. Carry through on the assembly process by applying clamps and tightening them. Check the box to see if it is closing square. Having established that the box does go together square and true, you can get ready to glue these parts together.

3. Remove the clamps and disassemble the chest.

4. Assemble the box walls using glue. Apply glue on just one of the four joints and close it. Then apply glue to a second joint and close

Photo D: You'll be least likely to damage the joints if you stand the parts on edge to assemble them. Once you have an end joined to the back, they'll be self-supporting, freeing both hands to maneuver the second end, then the front, into position.

Photo E: Nail the cleats for the box floor with their bottom edges flush with the drawer opening. Nail the cleats for the bottom with their edges flush with the edges of the box walls.

it. Next add glue to the last two joints, since both have to be closed at the same time. Get those joints closed and apply the clamps, just as you practiced it during dry-assembly. Check the piece with a square, and then let it alone to cure.

Attaching the cleats

With the glue cured and the box free of clamps, cut and fit the cleats that support the box floor and the bottom.

1. Rip the cleat stock to width. For the box floor, the cleats should be 1¼ in. wide. For the bottom, the cleats are 1½ in. wide.

2. Measure across the box, front to back, at each end.

3. Crosscut two cleats of each width to that length.

4. Glue and nail the cleats to the inside of the box. Because the grain in the cleats is parallel to that in the box walls, you can glue the parts together with impunity. Use cut nails, and drill pilot holes for each before driving the nail (see photo E).

5. Measure for and cut the long cleats. These fit between the end cleats and, like them, are glued and nailed to the chest walls. There are two long cleats for the box floor, but only one for the bottom. The front edge of the bottom rests on the front rail instead of on a cleat.

Fitting the drawer partition

The bottom slides into the drawer opening on the cleats, and its front edge is flush with the face of the front rail. The drawer partition splits the drawer opening in half. To have the access you need to install the bottom and the partition, you must do that before setting the box floor in place.

1. Measure the box for the bottom. You need the depth of the drawer opening from the front surface of the box to the inside back wall and the width from end to end. This is the area covered by the bottom.

2. Cut the bottom panel to these dimensions.

3. Trim the blank for the drawer partition to the final dimensions. Measure the box to be sure the divider dimensions specified by the cut list will work. (You want each drawer opening to be 21⅜ in. wide; adjust the thickness of the divider as necessary to achieve these measurements.) Mill the stock to the correct thickness, then rip and crosscut the blank. Lay out and cut the notches for the cleats that sit above it.

4. Lay out and cut the dado in the bottom for the partition. It should be an equal distance from each end. It begins at the back edge and stops about ½ in. from the front edge. Cut it ⅛ in. deep. I cut this on the router with a large-diameter straight bit, trapping the router between two guide fences to establish and control the width of the cut.

5. Slide the bottom in place and, as you do, capture the partition between it and the cleats above it (see **photo F**). Because the partition is cross-grain to the bottom, it's best not to glue it along its full length. Instead, glue just the 3 in. or 4 in. at the front end. The bottom can be glued to the front rail but not to the back or end cleats. You can drive several cut nails through the bottom into the end and back cleats.

Photo F: With the drawer partition clamped temporarily to the rear cleat, maneuver the bottom into the opening and slide it in, with the partition in the dado. Clamp it to the front rail before it seats, so you can apply glue to the underside of its front edge and the front of the dado. Remove the clamps to seat the bottom.

6. The partition is captured by the dado. Check with a square to ensure that it is at right angles to the bottom; then secure it by driving cut nails at an angle through the partition into the cleats. You can also drive a nail or two through the back into the partition.

Photo G: The till joinery is primitive but practical. Slide the glued-up bottom and back into place against the left end of the box, where it is secured with glue and a couple of cut nails. Then the front is glued into its groove in the bottom and toe-nailed to the box front and back.

Fitting the box floor

The box floor is cut to match the interior dimensions of the box, and it rests on the cleats. Quarter-round molding holds it in place. It is unrestrained and can expand and contract seasonally.

1. Measure the width and length of the box interior.

2. Rip and crosscut the box floor to those dimensions.

3. Drop the floor into place. If you want to, you can secure it with a few cut nails.

4. Cut the quarter-round molding. You can use stock molding or you can make your own by routing the profile on strips of the working stock using a table-mounted router. In either case, you should crosscut two strips of the molding to fit tightly from end to end and two strips to fit side-to-side against the box back. Both ends of all pieces are square-cut.

5. Glue and nail one of the long strips to the box back (but not the box floor). Make sure the molding is seated tight to the floor before driving the nails.

6. Cope one end of an end strip to mate with the already installed back strip.

7. Cut, cope, and fit the remaining to strips of molding. The first of these need be coped on one end only; but the second piece, the last one to be installed in the box, will have to be coped on both ends. Glue and nail them in place.

Making the till parts

The till is a small chest within the box located at the left end, ½ in. below the top edge. It also functions as a prop for the chest lid. You lift the lid, flip open the till lid, then lower the chest lid gently onto the smaller lid's corner.

The till is constructed of ¾-in.-thick stock. It has five parts, two of which have a profile routed on the front upper edge.

1. Mill the till stock to the specified thickness, if necessary, and rip it to width. Measure the interior width of the box before crosscutting the parts, just to be sure they will fit properly. You want the bottom, front, and hinge ledge to fit snugly between the box front and back. Crosscut the parts.

2. Profile the front edge of the till bottom and the till lid. Use the same profile on both. Smith used a cove-and-bead profile (often referred to by bit manufacturers as a "classical" profile).

3. Rout a ¾-in.-wide by ¼-in.-deep groove in the till bottom for the till front. The exact position isn't critical, but leave a shoulder about ⅛ in. wide between the profile and the groove.

Assembling the till

The till is basically assembled in place.

1. Glue the till bottom to the till back. Make sure the edges of the bottom are flush with the face of the back.

2. Mount the till in the box. I tipped the box onto its left end to do this (see **photo G**). Glue the till hinge ledge to the till back and the box side. Finally, hinge the till lid to the ledge.

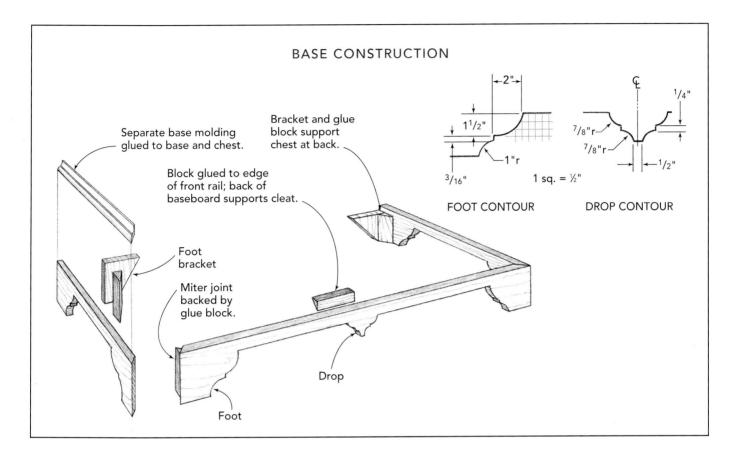

BASE CONSTRUCTION

Separate base molding glued to base and chest.

Bracket and glue block support chest at back.

Block glued to edge of front rail; back of baseboard supports cleat.

Foot bracket

Miter joint backed by glue block.

Drop

Foot

2"

1 1/2"

1"r

3/16"

1 sq. = 1/2"

FOOT CONTOUR

1/4"

7/8"r

7/8"r

1/2"

DROP CONTOUR

BUILDING THE BASE

The base consists of three baseboards, a pair of small brackets used as feet at the back, and several glue blocks, as shown in "Base Construction." The waist molding is separate.

Cutting the main parts

1. Dress the 1-in.-thick stock for the base, if you didn't do it earlier.

2. Rip the stock to 4¼ in. wide; then crosscut the front and end pieces to their rough lengths, or about 2 in. longer than specified.

Mitering the base pieces

Before laying out and sawing the foot cut-outs, you need to fit the pieces to the box. Doing this enables you to locate the cutouts accurately.

1. Fit the piece that will be at the front of the chest and mark it for mitering.

2. Miter the base pieces. The cuts are actually bevel cuts, produced with the work flat on the saw table and the blade tilted to 45 degrees for the cut. This is the case for both a table saw and a power miter saw. The end pieces are mitered on one end and cut square on the other. The front piece needs to be cut carefully so you get tight miter joints on each end.

Cutting the base contours

The feet are cut out of the three base pieces. The front of the base has an ornamental drop as well as feet at the ends.

1. Make a template following the pattern shown in "Base Construction" (above). Use a piece of cardboard and mark the arcs of the cove-and-bead contour with a compass or drafting templates. Cut out the cardboard pattern.

2. Use the pattern to lay out the cuts at each end of all three base pieces. Note that the bottom of the foot is 4 in. long on the front piece but only 3 in. long on the end boards. Draw a straight line with a rule to connect the foot contours, 2¹¹⁄₁₆ in. from the bottom edge.

Photo H: A sensible
strategy for cutting
the base contours
is to use a jigsaw
guided along a
clamped-on fence
to make the straight
cuts from foot to
drop. The curves
can be then cut
with the jigsaw or
on the bandsaw.

3. Lay out the drop on the front board. There's only one of these, so it may not be worthwhile to make a pattern. Follow the drawing.

4. Cut to the layout lines on the bandsaw or with a jigsaw (see **photo H**).

5. Sand the cut edges as well as the faces of the base pieces.

Assembling the base

The miter joints between the base pieces are glued together and then reinforced with glue blocks.

1. Apply glue and join the miter joints that connect the front to the end boards. Make sure the joints are square.

2. Cut the corner glue blocks, one for each miter joint. To avoid having a cross-grain glue-up here, crosscut the blocks from 2-in.-thick stock that's at least 3⁵⁄₁₆ in. wide. The crosscut should be a bevel cut, so you get a triangular block. Cut the blocks so they'll be 3⁵⁄₁₆ in. high.

3. Glue a block in the inside corner formed by each miter joint. Make sure the blocks are flush at the bottom of the assembly. The box will rest on these blocks.

4. Cut the rear foot brackets. These are pieces of the ¹³⁄₁₆ in. stock, ripped to 3⁵⁄₁₆ in. wide. Cut a piece 12 in. to 18 in. long. Miter each end; then crosscut a 4⁵⁄₁₆-in.-long piece from each end. This will provide you with two parts, each with 4⁵⁄₁₆-in.-long edge to glue against the edge of the box back and a 1-in.-long foot pad.

5. Cut corner glue blocks for the rear foot brackets. These glue blocks are the same as the front blocks, but cut from 1-in.-thick stock. At the same time, cut a block to reinforce the drop.

6. Glue the rear foot brackets and their glue blocks to the base. Glue the end of the foot bracket to the inner face of the end board. Glue the glue block into the inside corner.

Joining the base and box

The box should just fit inside the base and rest on top of the glue blocks and the rear foot brackets.

1. Turn the box upside down and test the fit of the base.

2. Apply glue to the inside surface of the base boards, from the glue blocks up to the top

Photo I: Set the base assembly onto the bottom of the up-ended box after spreading glue on the inside edges. Work it down over the box until the glue blocks are seated against the box's bottom edges. Then apply pipe or bar clamps.

edges. Use a brush to spread it evenly. Apply glue also to the top surfaces of the glue blocks and the foot brackets.

3. Set the base in place on the box (see **photo I**). Apply clamps from front to back and from end to end. The grain direction in the base pieces parallels that in the box surface to which it is being glued. It's all long-grain to long-grain, so the glue joint should be strong.

4. Glue the reinforcing block to the back of the drop and the front rail edge.

Routing and applying the base molding

The base molding is routed on a single board ⅞ in. thick by 3 in. wide by about 50 in. long. If you can find a single large bit to do the profile, use it. Failing that, here's a two-bit approach you can use with a table-mounted router.

1. Set up the router table with a ⅜-in. ovolo bit. This is basically a roundover bit without the pilot bearing. Lay the work flat on the table and feed it along the fence. Adjust the height of the bit and position the fence as shown in "Making the Base Molding" on p. 54. Cut this shape on both edges of the workpiece.

2. Switch to a corebox or round-nose bit to rout the cove. Again, set the bit height and the fence position as shown. To make this cut, you stand the work on edge and hold it firmly against the fence (see **photo J** on p. 54). Rout the cove into both edges.

3. Rip the molded edges from the workpiece. It's best to make the rips so the narrow molded edge falls free to the outside of the blade, rather than trying to feed the last few inches of the shaped piece between the rip fence and the blade.

4. Fit and cut the base molding strips. The molding is glued to the top edge of the base and the walls of the box. The front corner joints are mitered; at the back, the molding is cut flush with the box back. Divide one of the two molding strips in two, and miter one end of each piece. Fit the molding strips to the box, one by one, and glue them in place.

MAKING THE BASE MOLDING

Step 1. Rout the bead.

7/8"
board

7/16"

Fence

3/8"

Ovolo bit

Step 2. Rout the cove.

Workpiece

Fence

5/16"

Roundnose bit

Step 3. Saw off the molding.

13/16"

FINISHED MOLDING

1/8"

5/16"

13/16"

3/8"
bead

7/8"

Photo J: Routing the base molding profile is a two-step operation. Cut the quarter-round shape with a ovolo bit andcove with a round-nose bit. To make the cut, the workpiece is placed on edge and fed along the router table fence.

Applying the waist molding

While not strictly a part of the base, the waist molding shouldn't be attached until after the base is in place.

1. The waist molding profile is a ¼-in.-diameter bead milled on a ⅜-in. by ⅞-in. base. Make the cut with a ¼-in. bullnose bit in a table-mounted router. Cut it on a ⅞-in. by 2-in. or 3-in. strip of stock that's at least 50 in. long. Mill both long edges.
2. Rip the two molding strips.
3. Fit the molding to the chest, mark the pieces for the miter joints, and miter the ends.
4. Glue and nail the strips to the chest.

BUILDING THE LID

The chest lid is nothing more than a broad panel with a molding attached to three edges. Often on lids of this sort you'll see cleats attached to the underside to counter the tendency of a wide panel to cup. In this design, the lid molding is attached the way a breadboard end would be, and so it helps keep the lid flat.

Making the lid

1. If you didn't glue up the panel for the lid earlier, do so now. Next trim the lid to size and cut a tongue for the molding on three edges.
2. Rip and crosscut the lid to the dimensions specified. Make sure it is square.
3. Form a ¼-in.-thick by ½-in.-long tongue on the front and end edges of the lid. The tongue should be centered on the edges. You can make the tongue with a dado head in the table saw or with the router (see **photo K**).

Routing the lid molding

Like the base molding, you can make all the lid molding needed from a single 3-in.-wide board. It is less risky to mold the edge of a board this wide than it is to mold a 1-in.-wide strip.

1. Mill a piece of stock 1½ in. thick by 3 in. (or more) wide by about 52 in. long. It should be straight grained.
2. Use a molding bit to shape the wood. The particular bit you'll use has no common name, though it is a profile common to most bit manufacturers' catalogs (for example, CMT: #855.902.11; Amana: #54220; Freud: #99-015; and Jesada: #655-902). Do the cut on the router table. Mill the profile on both edges of the stock.
3. Rip the molded edges from the workpiece, cutting each ¹⁵⁄₁₆ in. wide.
4. Cut a groove in the back of the molding for the tongue you cut on the lid's edges. The groove must match the size of the tongue, and it must be properly positioned so the molding is flush with the top surface of the

Photo K: You can form the tongue on the lid's edge with a router. I used a large-diameter mortising bit for the cut, guiding the router with an edge guide. Making a final pass along the three edges of both sides will center the tongue on the edge.

Tip: Mounting the waist molding doesn't require three hands. Tip the chest onto its back, and locate the molding with two or three spacers, referencing off the base molding. Then "clamp" it to the chest with masking tape.

Photo L: The lid molding fits onto the tongue routed in the chest lid, fitting flush with the top surface and overhanging the bottom. It is substantial enough to act as a breadboard end for the lid, keeping it from bowing.

Tip: To cut the groove, set up whichever machine you elect to use and make a test cut on pieces of the working stock. Test-fit the grooved piece to the lid. You want a snug but not tight fit. If necessary, adjust the machine setup and cut a new test piece.

lid. The groove can be cut on the table saw or on the router table. Make a test cut first; once you get the machine set up, cut the groove in the stock.

Attach the molding

As with a breadboard end, some care must be taken *not* to glue the moldings across the ends of the chest lid. These moldings must be partially glued and pinned so the lid can expand and contract.

1. Miter the front molding as a starting point. Fit the uncut molding onto the tongue along the lid's front and mark it for the miter cuts. (see **photo L**). Miter the molding.
2. Glue the front molding to the lid.
3. Fit the end moldings to the lid. Divide the remaining long molding strip into two, and miter one end of each piece (don't miter the same end of each, though). Fit the moldings to the lid without glue. Make sure you have

nice, tight miters. Now mark two or three spots on the molding for assembly pins.
4. Drill pilot holes for the assembly pins. Plan to use ⅛-in. dowel for the pins. So drill a ⅛-in.-diameter hole at each of your marks, penetrating through the top of the molding and the tongue and into—but not through—the section of the molding beneath the tongue. The holes should be centered in the tongue.
5. Remove the molding and elongate the pilot holes in the tongue. You can do this with a ⅛-in. straight bit in the router. The idea is to lengthen the holes parallel to the end of the tongue (across the grain).
6. Attach the end moldings. Spread glue on the miter and along the molding (and in its groove) extending no farther than 3 in. or 4 in. away from the miter. Fit the molding onto the tongue, press the miter joint tight, and clamp the molding. Then drive the dowels into their holes. When the glue has set, remove the clamps and pare the dowels flush.

Hinging the lid

The lid is hinged to the chest with two iron strap hinges. Smith buys hinges from David Fisher, a blacksmith in Hamburg, Pennsylvania, but appropriate hinges are available elsewhere (see "Sources" on p. 216).

1. Lay out the locations of the hinges. Smith mounted them about 7½ in. from the ends of the chest. The leaf that attaches to the box is L-shaped. The hinge barrel is outside the chest, and the L-shaped leaf extends across the top edge of the box back and down its inner face. You probably will need to notch the top edge of the box back to accommodate the hinge leaf and have the lid rest flat atop the box. Lay out these notches and determine how deep they must be.
2. Cut the notches.
3. Screw the hinges to the chest. The mounting-screw holes are in the leaf extension that seats against the inner face of the box back. Drill pilot holes and drive the screws.
4. Close the hinges and lay the lid in place. Mark the hinge locations on the back edge of the lid. Remove the lid and the hinges from the chest.
5. Screw the hinges to the lid. Using the marks on the lid's back edge, line up the hinges on the inner face of the lid. Drill pilot holes and drive the mounting screws.
6. Mount the lid on the chest. Hold the lid upright and lower it to set the hinges in the notches. Redrive the mounting screws. Remember that the till lid is the prop that holds the chest lid open.

Mounting the lock

If you'd like to lock the chest, install a lock, as Smith did. You need to have it in hand before cutting the mortise or drilling the keyhole.

1. Determine the dimensions of the mortise for the lock and its relation to the keyhole. You want the keyhole to be centered on the chest front.
2. Lay out and drill the keyhole.

3. Lay out the cut the mortise. You can use a router to hog out the majority of the waste; then pare to the layout lines and square the corners with a chisel. Check the fit of the lock in the mortise and then adjust the mortise as necessary.
4. Use the lock as a pattern to lay out the mounting-screw holes. Drill the holes and mount the lock, as a sort of dry assembly.
5. Establish the position of the catch. Mark the lid; then lay out and drill the mounting-screw holes for the catch. Screw it in place.
6. Check the latching and/or locking action. Make sure it works as it is supposed to. Then remove the lock and the catch. Set them aside until the chest is painted.

MAKING THE DRAWERS

This chest has but two drawers, and they are identical. The drawer fronts are lipped all around and have Chippendale-style brass pulls and matching keyhole escutcheons. The sides and front are joined with half-blind dovetails; the sides and back are joined with through dovetails. The bottom is solid wood.

Cutting and fitting the parts

1. Dress the stock for the drawers, if you haven't already done so. Mill the stock for the sides, backs, and bottoms to ⅝ in. thick. The fronts are ¹³⁄₁₆ in. thick.
2. Glue up narrow boards to make the rough-size panels needed for the two bottoms.
3. Measure the height, width, and depth of both drawer openings.
4. Rip and crosscut the drawer sides to size.
5. Rip and crosscut the drawer fronts to size. Bear in mind that the fronts are rabbeted all around. You want the fronts' tops, bottoms, and ends to overlay the boxes and drawer partitions, but you still need clearance between the boxes and the shoulders of the rabbets so the drawers don't stick.
6. Rabbet the drawer fronts. The cut is ⅜ in. by ½ in. along the bottom and across the ends, but ½ in. by ½ in. along the top. Choose the best face, mark it, and cut the rabbets into the

DRAWER CONSTRUCTION

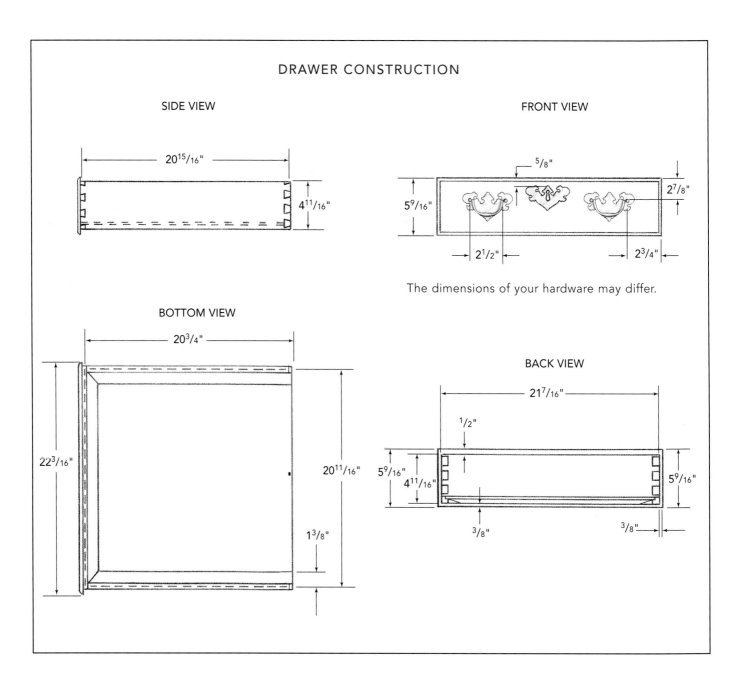

SIDE VIEW

20¹⁵/₁₆"

4¹¹/₁₆"

FRONT VIEW

5/8"

5⁹/₁₆"

2⁷/₈"

2¹/₂"

2³/₄"

The dimensions of your hardware may differ.

BOTTOM VIEW

20³/₄"

22³/₁₆"

20¹¹/₁₆"

1³/₈"

BACK VIEW

21⁷/₁₆"

1/2"

5⁹/₁₆"

4¹¹/₁₆"

5⁹/₁₆"

3/8"

3/8"

opposite one. Fit the fronts to their respective openings.

7. Finally, rip and crosscut the backs. The backs will be the same length as the fronts' shoulder-to-shoulder dimension. They will be ¹¹/₁₆ in. narrower than the sides.

Cutting the joinery

1. Dovetail the sides and backs. These are through dovetails. The top edges are flush; the back is recessed ¹¹/₁₆ in. so the drawer bottom can slide beneath it. The joints can be laid out and cut by hand or with a router and dovetailing jig.

2. Dovetail the fronts and sides. These are half-blind dovetails

3. Groove the sides and front for the bottom. The groove is ¼ in. wide by ¼ in. deep and the top edge of it is ¹¹/₁₆ in. from the bottom edge of the parts. The grooves can be cut on a router table or on the table saw.

Drilling for the brasses

While you aren't going to mount the brasses now, this is a good time to drill the mounting holes for them. *Brasses* is a collective name for the bail-type pulls with decorative mounting plates and the matching keyhole escutcheons used on eighteenth- and nineteenth-century furniture. Reproductions of the old brasses are available from a number of sources. Smith got the brasses for this chest from Horton Brasses.

1. Determine where you want the drawer pulls to be on the drawer fronts. The positioning used by Smith is shown in the drawing on the facing page. Lay out a pair of pulls and find the best side-to-side positioning.
2. Determine where the mounting holes must be located and lay them out on the drawer fronts.
3. Bore the mounting holes on the drill press.
4. Lay out the location of the keyhole escutcheon. Smith placed the tip of this plate on the vertical centerline of the drawer front and $\frac{5}{8}$ in. down from the edge. Mark this spot on each drawer front; and from that point, lay out the mounting hole positions.
5. Drill pilot holes for the nails that attach the escutcheons.

Mortising for the locks

If you'd like, install a lock in each drawer.

1. Working from the escutcheon location, lay out the mortise required on the inside of the drawer front.
2. Establish how deep the mortise must be.
3. Rout out most of the waste. You can cut freehand, working close to the layout lines, or you can make a template to guide the router, thus eliminating the chance you'll inadvertently wander past the lines.
4. Square the corners of the mortise with a chisel. Pare to the lines.
5. Set the lock into place to ensure that it fits. Transfer mounting-screw locations to the wood and drill pilot holes. It's a good idea to screw the lock in place to make sure it fits; then remove it and set it aside until after the chest is completed and has been finished.

Assembling the drawers

With the fronts rabbeted and the dovetails cut, the drawers are all but ready for assembly. You need to rout the round-over profile on the fronts. And the bottoms must be beveled to reduce the edge thickness.

1. Rout the profile on the drawer fronts. This is a good router-table operation. Do the end-grain cuts first and provide good backup so you don't get tearout. Then make the long-grain cuts.
2. Bevel the bottoms. Although the bottoms are $\frac{5}{8}$ in. thick, the grooves for them are $\frac{1}{4}$ in. wide. The edges of the bottom need to be beveled to reduce the edge thickness. It is like raising a panel. You can handplane the edges or rout them with a straight-bevel panel-raising bit. If you do the latter, you'll get a nice tongue to fit the grooves.
3. Notch the back edges of the bottoms. To secure the drawer bottom, you drive a fastener or two through it into the drawer back. Because the bottom will expand and contract, some wiggle room around the fasteners is important. The notches, just $\frac{1}{8}$ in. or so wide and $\frac{1}{2}$ in. long, provide that.
4. Dry-assemble each drawer and fit it to the case. This provides a last chance to rectify fit problems before the drawer is glued up.
5. Using the glue now, assemble each drawer. Join the sides to the front and back, making sure the assembly is square and rests flat. Slide the bottom into the grooves. After drilling pilot holes into the back, drive the nails to secure the bottom.

FINISHING UP

Smith uses oil-base paints exclusively. The drying times are fairly protracted. He has a separate finishing room, one that's actually in a separate building, so he can be making sawdust while the paint dries. Here, step-by-broad step is how Smith applies the finish. Note that the drawer fronts and lid are painted at the same time the chest is.

Photo M: The first coats of paint go on the chest only after the moldings and the panels have been masked off. Apply two coats of paint over the sealer and let both dry thoroughly.

1. Remove the lid and the drawers, so they can be painted separately. Set the chest on a low work table, so you can sit beside it and have the vertical surfaces at eye level.

2. First apply a clear sanding sealer, covering the chest inside and out.

3. When the sealer is dry, outline the three panels on the chest front. Mask these off with blue painter's masking tape, which won't pull the paint off. In addition, mask off the trim: the waist molding and the base molding.

4. Two coats of oil-base paint are applied to the main parts of the chest (those parts not masked off) (see **photo M**).

5. When the main body is done, strip off the masking tape. Then mask off the body. Next paint the panels in the solid background color and then paint the trim. Apply two coats to these areas.

6. Next, lay out the urns and figures. Especially where symmetry is important—the unicorns on the center panel, the urns, the distelfinks (the stylized bird characteristic of Pennsylvania Dutch decorations), the

Photo N: Poster-board patterns can be used to lay out the features in the painting that need to be symmetrical.

Photo O: The most elaborate decorations have to begin somewhere. Smith paints the elements laid out from the patterns before progressing to the spreading vines.

horsemen—use patterns to facilitate your layout (see **photo N**).

7. Next paint in the laid-out elements. Approach this very methodically. Draw everything in pencil before you apply the paint. Note that Smith doesn't draw it in all at once. And most of the drawing is done freehand, as is all of the painting (see **photo O**).

8. When the painting is completed, apply two coats of satin varnish.

9. When the varnish is dry, install the locks, escutcheons, and drawer pulls.

CONTEMPORARY CHEST

Think of a "chest of drawers." The mental image I get is a chest very much like this: two half-width drawers over three or four full-width ones in a case about 42 in. high and 36 in. wide. I've seen this form again and again in all sorts of styles, from Jacobean to William and Mary to Queen Anne, through Arts and Crafts and Shaker, right up to the most contemporary of styles.

This particular example, though a slight departure from some of the details of the form, nevertheless captures the essence of it perfectly. The number and the sizes of its drawers are just right for a person's wardrobe. You have a place for everything you don't have on hangers in the closet. The drawers are manageable sizes, and there's a range of them. You have a deep drawer or two for bulky trousers and sweaters and shallow ones for your tanks, boxers, and socks. The size of the case is just right too for any but the smallest bedrooms and all but the smallest of users.

Designed and built by Ken Burton, this chest offers a contemporary look and sound, modern construction. The cherry is clear but not highly figured, so premium stock isn't essential. The lines include tapers and angles that banish the boxiness without being difficult to duplicate. The construction procedures are tailored to the experienced hobby woodworker with a reasonably well-equipped shop.

Contemporary Chest

DON'T BE DAUNTED by the dozens of loose tenons used in assembling this chest of drawers. The mortises that house those tenons are routed in just five setups, right at the outset of the construction. In short order, the framework of the chest is ready for assembly. Note that the bottom drawer front is shaped to match the recess in the bottom front drawer rail.

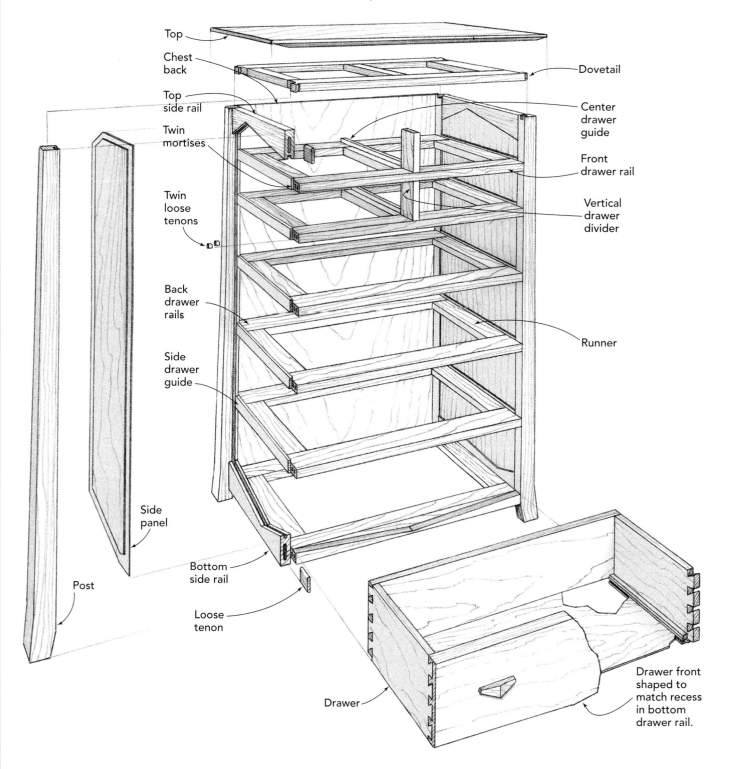

Top

Chest back

Top side rail

Twin mortises

Twin loose tenons

Back drawer rails

Side drawer guide

Side panel

Post

Bottom side rail

Loose tenon

Drawer

Dovetail

Center drawer guide

Front drawer rail

Vertical drawer divider

Runner

Drawer front shaped to match recess in bottom drawer rail.

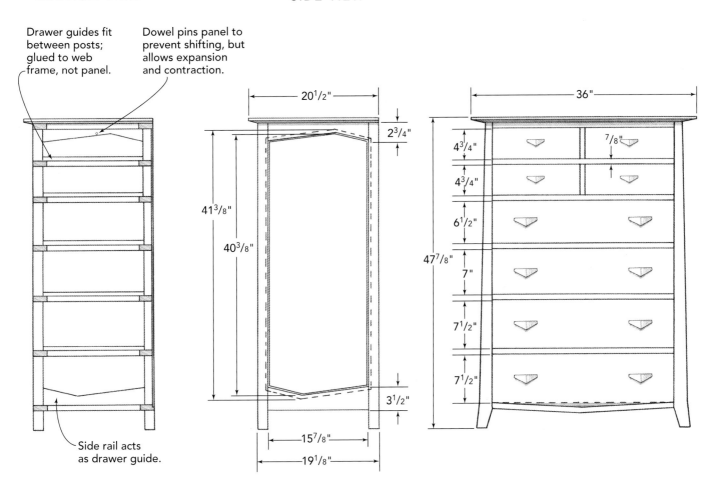

SECTION VIEW

Drawer guides fit between posts; glued to web frame, not panel.

Dowel pins panel to prevent shifting, but allows expansion and contraction.

Side rail acts as drawer guide.

SIDE VIEW

20 1/2"

2 3/4"

41 3/8"

40 3/8"

3 1/2"

15 7/8"

19 1/8"

FRONT VIEW

36"

4 3/4"

7/8"

4 3/4"

6 1/2"

47 7/8"

7"

7 1/2"

7 1/2"

The bottom side rails act as guides for the bottom drawer. The guides for the other drawers fit between the posts and are glued to the web frame (not the panel).

BUILDING THE CHEST STEP-BY-STEP

CUT LIST FOR CONTEMPORARY CHEST

Case			
4	Posts	1⅝ in. x 2½ in. x 47⅛ in.	cherry
2	Top side rails	⅞ in. x 2¾ in. x 15⅞ in.	cherry
2	Bottom side rails	⅞ in. x 3½ in. x 15⅞ in.	cherry
1	Top front drawer rail	⅞ in. x 2¼ in. x 30¼ in.	cherry
1	Top back drawer rail	⅞ in. x 2 in. x 30¼ in.	cherry
1	Bottom front drawer rail	1⅝ in. x 2¼ in. x 29 in.	cherry
5	Front drawer rails	⅞ in. x 2¼ in. x 29 in.	cherry
6	Back drawer rails	⅞ in. x 2 in. x 29 in.	cherry
2	Side panels	9⁄16 in. x 16⅝ in. x 41⅜ in.	cherry
2	Loose tenons	5⁄16 in. x 1⅛ in. x 1½ in.	cherry
2	Loose tenons	5⁄16 in. x 2 in. x 1½ in.	cherry
34	Loose tenons	5⁄16 in. x 1¾ in. x 1½ in.	cherry
4	Loose tenons	5⁄16 in. x 1½ in. x ¾ in.	cherry
48	Loose tenons	5⁄16 in. x ½ in. x 1½ in.	cherry
17	Drawer runners	⅞ in. x 2¼ in. x 14⅝ in.	cherry
2	Vertical drawer dividers	¾ in. x 2¼ in. x 4¾ in.	cherry
10	Drawer guides	¼ in. x 1½ in. x 15⅞ in.	cherry
2	Drawer guides	¾ in. x ¾ in. x 16 in.	cherry
1	Top	⅞ in. x 20½ in. x 36 in.	cherry
1	Back	¼ in. x 29½ in. x 43⅝ in.	birch plywood

MY MENTAL PICTURE of this chest during construction is of a blizzard, a swirl of loose tenons, little flakes of wood drifting in the air. The chest is held together with almost eight dozen of them—four per side assembly, four to six per web frame, eight miniatures joining web frame to posts, and a pair per vertical drawer divider. But the dominant process, once the stock is dressed and the key parts are cut to size, is not making the tenons themselves, but cutting the 180 mortises that house those tenons.

Begin the project by dressing the stock. While the boards acclimate themselves to your shop's environment in stickered stacks, consider the mortising options. The first real work is cutting those mortises.

CUT LIST FOR CONTEMPORARY CHEST

Drawers

4	Drawer fronts	¾ in. x 4¹¹⁄₁₆ in. x 14⅛ in.	cherry
1	Drawer front	¾ in. x 6⁷⁄₁₆ in. x 29 in.	cherry
1	Drawer front	¾ in. x 6⅞ in. x 29 in.	cherry
1	Drawer front	¾ in. x 7⅜ in. x 29 in.	cherry
1	Drawer front	¾ in. x 8⅛ in. x 29 in.	cherry
8	Drawer sides	½ in. x 4¹¹⁄₁₆ in. x 18⁹⁄₁₆ in.	maple
2	Drawer sides	½ in. x 6⁷⁄₁₆ in. x 18⁹⁄₁₆ in.	maple
2	Drawer sides	½ in. x 6⅞ in. x 18⁹⁄₁₆ in.	maple
4	Drawer sides	½ in. x 7⅜ in. x 18⁹⁄₁₆ in.	maple
4	Drawer backs	½ in. x 4³⁄₁₆ in. x 14⅛ in.	maple
1	Drawer back	½ in. x 5¹⁵⁄₁₆ in. x 29 in.	maple
1	Drawer back	½ in. x 6⅜ in. x 29 in.	maple
2	Drawer backs	½ in. x 6⅞ in. x 29 in.	maple
4	Drawer bottoms	¼ in. x 18⅜ in. x 12⅝ in.	birch plywood
4	Drawer bottoms	¼ in. x 18⅜ in. x 27½ in.	birch plywood
16	Drawer slips	½ in. x ¾ in. x 18⅜ in.	maple
4	Small drawer pulls	¹¹⁄₁₆ in. x 1 in. x 3 in.	cherry
8	Large drawer pulls	¹³⁄₁₆ in. x 1¼ in. x 3½ in.	cherry

All case parts are cherry—except the back, which is birch plywood. The drawer fronts are cherry, the drawer sides and backs are maple, and the drawer bottoms are birch plywood.

BUILDING THE SIDE ASSEMBLIES

Right at the start, you need to prepare virtually all the case parts, except the top. The side assemblies and web frames have the mortises, almost all of which you need to cut before you assemble anything. That done, you focus on completing the side assemblies.

Each side assembly consists of two shaped posts, top and bottom side rails, and a side panel. The rails and posts are joined with loose tenons. The panel is captured in grooves in the rails and posts.

Preparing the stock

Loose-tenon joinery requires identical mortises to be cut in the mating parts. You want to cut all the matching mortises at the same time. So you need to have both the posts and all the rails ready to be mortised.

1. Lay out the posts, side rails, all drawer rails, and the side panels on rough stock

2. Rip and crosscut the parts to rough dimensions.

3. Joint a face and an edge of each part.

4. For each part, plane the second face flat and parallel to first and reduce the thickness to the final dimension.

5. Rip each piece to its final width plus ¹⁄₁₆ in.

MAKING A POST

Rout twin mortises for the web frames. Note that the top web frame is dovetailed into the post after the initial case assembly.

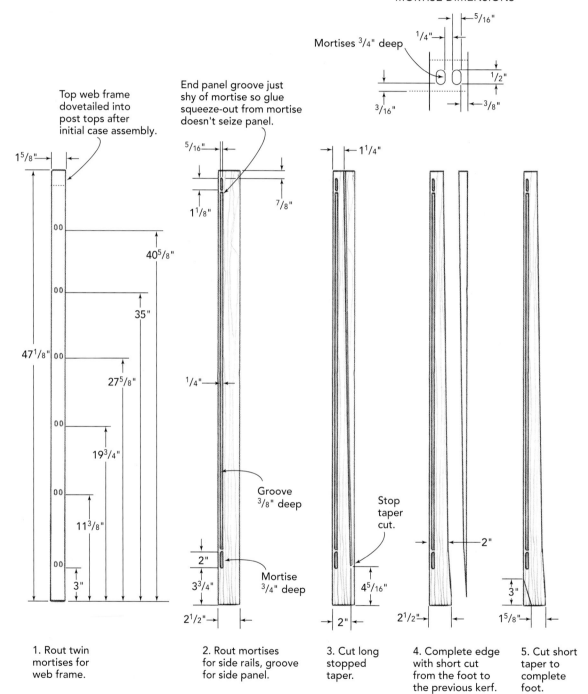

MORTISE DIMENSIONS

Mortises 3/4" deep

5/16"

1/4"

1/2"

3/16"

3/8"

Top web frame dovetailed into post tops after initial case assembly.

End panel groove just shy of mortise so glue squeeze-out from mortise doesn't seize panel.

1 5/8"

5/16"

1 1/4"

40 5/8"

1 1/8"

7/8"

35"

47 1/8"

27 5/8"

1/4"

19 3/4"

Groove 3/8" deep

Stop taper cut.

11 3/8"

2"

3/4"

3"

2"

4 5/16"

2"

2"

3 3/4"

Mortise 3/4" deep

2 1/2"

2"

2 1/2"

3"

1 5/8"

1. Rout twin mortises for web frame.

2. Rout mortises for side rails, groove for side panel.

3. Cut long stopped taper.

4. Complete edge with short cut from the foot to the previous kerf.

5. Cut short taper to complete foot.

6. Joint each part to the final width.

7. If necessary, edge-glue boards to form the side panels.

8. Cut all the posts and rails to their final lengths.

Routing the posts and rails

The mortises for joining the side and drawer rails are cut in the posts before the posts are shaped. At the same time you mortise the posts, do the mortises in all the rails. Burton cuts the mortises using a plunge router equipped with an edge guide and jig (see "Mortising Jig Plan" on p. 125). The jig has two interchangeable work holders: a horizontal one for posts and a vertical one for the rails.

1. Lay out the all the mortises on the posts, including the side rail mortises and the drawer rail mortises (see "Making a Post"). Lay the posts out together, so you get rights and lefts, fronts and backs and their mortises in the correct locations (see **photo A**).

2. Set up the work holder for routing the side rail mortises in the posts.

3. Set up the plunge router for cutting the side rail mortises. Use a ⁵⁄₁₆-in.-diameter straight bit in the collet. Set the maximum depth of cut to 1¼ in. Adjust the edge guide and the stops on the mortising jig.

4. Rout the mortises in the posts.

5. Switch work holders so you can rout the matching end mortises in the side rails. Align the work holder so the centerline of the mortise lines up with the jig's registration line. Adjust the router's edge guide so the mortise in the rail will be in the proper place.

6. Rout the mortises in the side rails.

7. Switch back to the horizontal work holder so you can cut the twin mortises in the posts. These mortises are cut with a ⁵⁄₁₆-in.-diameter straight bit. Adjust the edge guide to position the bit for cutting the far mortise. Set the stops for the short mortise. Burton routs the far mortise, then screws a spacer to the edge of the jig that the edge guide bears against. This shifts the router closer to the face of the jig, positioning the bit for routing the near mortise.

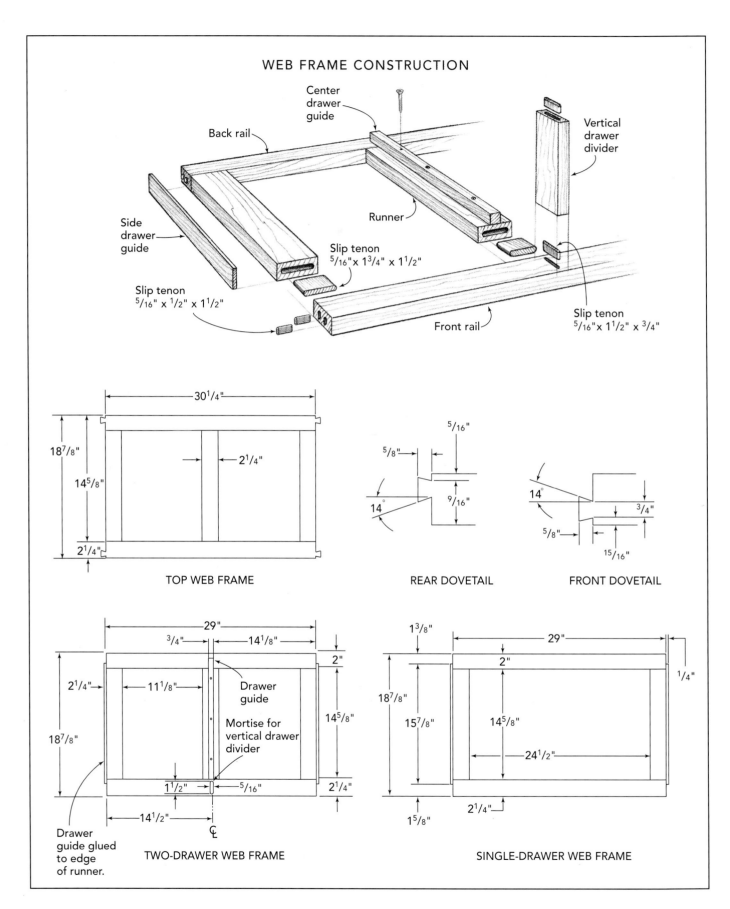

WEB FRAME CONSTRUCTION

Center drawer guide

Back rail

Vertical drawer divider

Side drawer guide

Runner

Slip tenon
$5/16$" x $1^3/4$" x $1^1/2$"

Slip tenon
$5/16$" x $1/2$" x $1^1/2$"

Front rail

Slip tenon
$5/16$" x $1^1/2$" x $3/4$"

TOP WEB FRAME

$30^1/4$"

$18^7/8$"

$14^5/8$"

$2^1/4$"

$2^1/4$"

REAR DOVETAIL

$5/16$"

$5/8$"

$14°$

$9/16$"

FRONT DOVETAIL

$14°$

$3/4$"

$5/8$"

$15/16$"

TWO-DRAWER WEB FRAME

29"

$3/4$"

$14^1/8$"

2"

$2^1/4$"

$11^1/8$"

Drawer guide

Mortise for vertical drawer divider

$14^5/8$"

$18^7/8$"

$1^1/2$"

$5/16$"

$2^1/4$"

$14^1/2$"

Drawer guide glued to edge of runner.

SINGLE-DRAWER WEB FRAME

$1^3/8$"

29"

2"

$1/4$"

$18^7/8$"

$15^7/8$"

$14^5/8$"

$24^1/2$"

$2^1/4$"

$1^5/8$"

Photo B: The post is aligned in the fixture and clamped to its face. When cutting the mortise, use the edge guide to reference the fixture's back edge; the stops limit the router's travel, thus controlling the length of the mortises.

Photo C: Screw a spacer to the fixture's back edge to shift the position of the bit ⅝ in. closer to the fixture. This allows you to rout both mortises in a pair with one router/edge-guide setup.

Tip: To halve the number of times he has to mount and remove the spacer, Burton alternates spacer and workpiece change-overs. He routs the far mortise in the first piece, then he mounts the spacer and routs the near mortise. He switches the workpiece and routs the near mortise. Then he removes the spacer and routs the far mortise. Continue the alternation until all the mortises are cut.

8. Rout the mortises in the posts (see **photos B** and **C**).

9. Switch work holders to do the matching twin mortises in the rail ends. Note that the short twin mortises are cut across the thickness of the rails rather than parallel to the width, as was the case with the side rails (see "Web Frame Construction"). Because the rails and posts are flush when assembled, the mortises are the same distance from the reference edge of both the post and rail. Thus, the router's edge guide doesn't need to be adjusted.

10. Rout the mortises in the rails (see **photo D** on p. 72). Be careful when mortising the bottom rail, which is thicker than the others. The mortises must be located in relation to the top surface. On one end, this is no particular problem. Before mortising the other, you need to shift the position of the work holder so the centerline will align with the jig's register line.

Cutting the grooves and rabbets

The last two joinery cuts to make in the posts are the groove for the side panels and, in the two back posts, the rabbet for the back panel.

Photo D: Switch work holders to use the fixture for mortising the rails. It shouldn't be necessary to alter the stops or the router/edge-guide setup. The mortises in the rails will match those in the posts.

The groove is ¼ in. wide and ⅜ in. deep and is ⁵⁄₁₆ in. from the post's edge. The rabbet is ¼ in. wide and ⁹⁄₃₂ in. deep (so the back panel is ever so slightly recessed).

1. Dog a post to the benchtop.
2. Set up a plunge router with a ¼-in.-diameter straight bit and an edge guide. Adjust the guide so the cut will be properly located in relation to the post's edge.
3. Rout the groove. To avoid the panel being inadvertently glued into the groove, stop the groove just shy of the side rail mortise on each end. Running the groove into the mortise invites glue to escape from the mortise onto the panel.
4. Set up a router either with a rabbeting bit or a straight bit and an edge guide.
5. Rout the rabbet in the inside back edge of the two back posts only. The rabbet extends 43⅜ in. from the top of the post, stopping where the bottom back rail joins the post.
6. Square the inside corner of the rabbet with a chisel.

Shaping the posts

The outside edge of the posts taper gradually from the top to the ankle, where the foot angles out sharply. The face and back are square, as is the inside edge, to which all the drawer rails join. The tapers are completed in three cuts.

1. Lay out the tapers on one post as shown in "Making a Post" on p. 68.
2. Make two tapering jigs to use on the table saw (see "Tapering Jigs"). Using the jigs will ensure you get consistent cuts from post to post.
3. Make the first cut on the table saw. This is a stopped cut that begins at the top of the post and extends only to the ankle, from which the foot cants away. Make the cut and stop short of the ankle (see **photo E** on p. 74). Be sure to account for the curve of the blade when establishing the stop point.
4. Turn off the saw. When the blade stops, lift the post and the jig off the blade. Don't mess with the waste. It will come off soon enough.

TAPERING JIGS

These two job-specific jigs are easy to make, do the job perfectly, and can be recycled without remorse when the posts have been tapered. Lay out the cut you want to make on the workpiece, align the work on a plywood or medium-density fiberboard (MDF) base, and screw a few fences to the base to hold the workpiece in position. Slide the base along the table saw's rip fence to cut the taper.

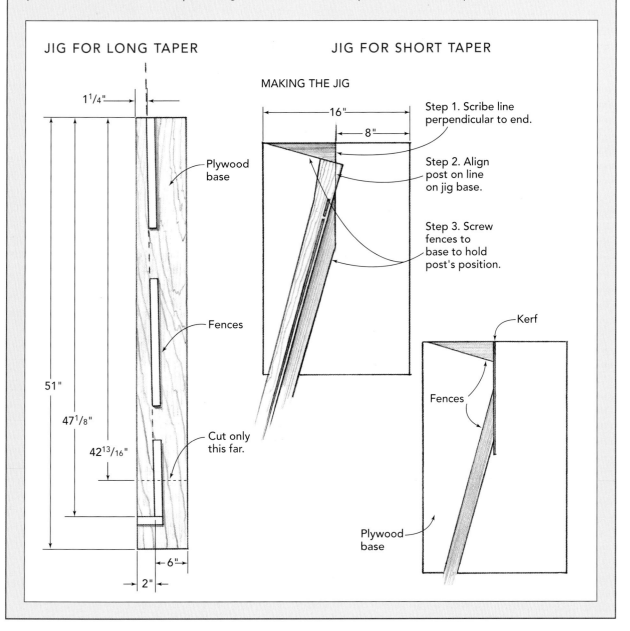

JIG FOR LONG TAPER

1¹/₄"

Plywood base

Fences

51"

47¹/₈"

42¹³/₁₆"

Cut only this far.

6"

2"

JIG FOR SHORT TAPER

MAKING THE JIG

16"

8"

Step 1. Scribe line perpendicular to end.

Step 2. Align post on line on jig base.

Step 3. Screw fences to base to hold post's position.

Kerf

Fences

Plywood base

Photo E: Cut the post's long taper on the table saw, guided by a tapering jig. Stop the cut when the blade reaches the ankle, the point at which the foot splays out sharply.

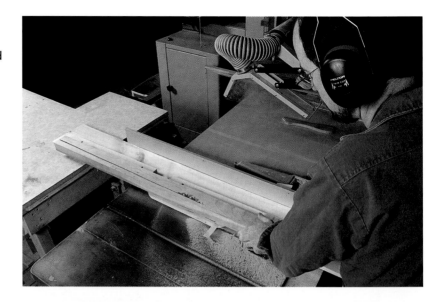

Photo F: A second tapering jig is used to cut the inner face of the foot. To save the jig, stop feeding the work when the waste is severed from the post. Then stop the saw and back the post off the blade.

5. Complete the outside post shape on the bandsaw. Cut from the foot in to the ankle. This is a short cut; and as it is completed, the long piece of waste will separate from the post.
6. Cut the inner edge of the foot on the table saw using the second tapering jig. The cut is stopped only to preserve the jig (see **photo F**).
7. Smooth the cut surfaces. Remove the ridge of waste left at the end of the long taper cut. Use a broad chisel. Then shave off the saw marks with a spokeshave or scraper. Sand all four surfaces of each post.

8. Use a block plane to chamfer the perimeter of the foot. This gives the appearance of floating just above the floor and will prevent chipping if the chest is dragged across a floor.

Making the side rails

The side rails have been cut to width and length, and they've been mortised. Now shape them and groove them for the side panels.

1. Lay out and cut the template for shaping the rails (see "Side Assembly Construction"). The shape is an asymmetrical V. Use the same template to shape both top and bottom rails.

SIDE ASSEMBLY CONSTRUCTION

Note that the top rails have the peak toward the back;
the bottom rails are the reverse.

PANEL CONTOUR

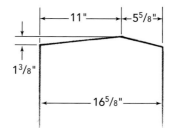

SIDE PANEL LAYOUT

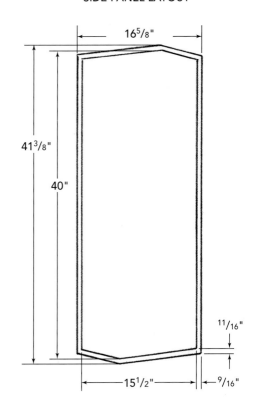

PANEL SHOULDER

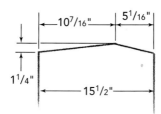

TOP SIDE RAIL LAYOUT

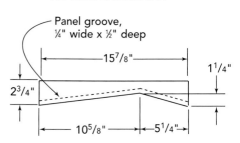

BOTTOM SIDE RAIL LAYOUT

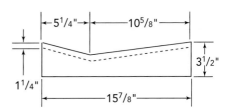

SIDE RAIL TEMPLATE

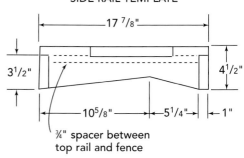

*Tip: Using tem-
plates can help
you produce consis-
tent, accurate cuts.
Make the templates
from thin plywood
or hardboard and
use them to lay out
the cuts. Use a
router and pattern
bit to trim the
workpiece flush
to the template.*

2. Cut the four rails with a router and pattern
bit that has a pilot bearing on the shank. This
can be done on a router table.
3. Cut the panel groove in the shaped rail
edge with a router and slot cutter.

Making the side panels

The 9/16-in.-thick side panel is raised, with the
raised field designed to be flush with the faces
of the side rails. The contour of the panel—
and of the raised field—matches that of the
frame formed by the posts and rails.

1. Cut the panels to their final dimensions.
2. Cut the panel contours. Note that each
panel is asymmetrical; don't lay out both ends
of a panel the same. The cuts can be made
with a jigsaw guided by a clamped-on fence.
3. Raise the panels. The shoulder of the panel
is square and the tongue is flat (see **photo G**).
You want a 3/16-in.-wide gap between the posts
and rails and the panel's shoulders. The
grooves in the rails are deeper—by 1/8 in.—
than those in the posts, so the width of cut
from edge to shoulder must be 1/8 in. wider

across the panel ends than along the edges.
You can make the cuts on the router table,
guided by the fence. With a handheld router,
make the cuts using a pattern bit guided by a
clamped-on fence or template.
4. Fit the tongues, produced by raising the
panels, into the post and rail grooves. Adjust
the fit by handplaning the tongues.
5. Scrape and sand the panels to ready them
for finishing and assembly.
6. Apply your finish to the panels. Burton
used an oil finish. Apply it to both sides and
all edges equally. The finish seals the panels
and ensures that expansion and contraction of
the wood will be consistent throughout the
panel. It also makes assembly a bit easier.

Assembling the sides

At this point, parts for both side assemblies are
ready to be joined.

1. Make the loose tenons needed for side
assemblies and fit them to the mortises. You
thickness and rip strips of stock, bullnose the
edges to match the radius of the mortise ends,

then crosscut the strips. You should be able to get a perfect fit with very little effort.

2. First assemble the units without glue (see **photo H**). Don't take this process lightly; it's important. It can steer you around the extraordinary pain and suffering resulting from a glue-up botched by a joint that doesn't quite fit. Think through the glue application, work out the best sequence of assembly, and set up your clamps.

3. Glue up the assemblies. Remember to keep the glue out of the panel grooves, so the panel can expand and contract. Don't worry too much about the alignment of the panel during the glue-up. Since it isn't glued, you can shift it later.

4. After the glue has cured and you've removed the clamps, align the panel so that gap between its shoulders and the posts and rails is consistent. If the gap is more or less than 3/16 in. at this point, you can't do anything about it; but you can line up the shoulders

parallel to the posts and make the gap on one side match that on the other.

5. Drill a hole through the inner face of the top rail, equidistant from the posts, into the panel's tongue. Glue a dowel into the hole, pinning the panel in the center. Repeat for the bottom rail. The panel will be able to move in and out from this fixed centerline.

MAKING THE WEB FRAMES

The drawers are supported by web frames, consisting of front and rear drawer rails and two or three runners. The front rails are made of cherry, the runners and rear rails of cherry or maple. The runners are joined to the rails with glued mortise-and-loose-tenon joints. The frames are joined to the side assemblies with twin mortise-and-tenon joints. These latter mortises have already been routed.

MAKING THE BOTTOM DRAWER RAIL

Cut the twin mortises before shaping the rail. The rail's top surface is flat to support the drawer; its bottom is tapered to parallel the contour. The drawer front will be contoured to match bottom of the recess.

BOTTOM DRAWER RAIL TEMPLATE

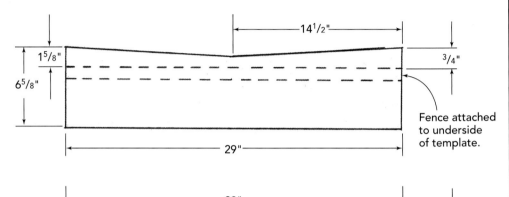

1. Cut blank.

2. Rout recess guided by template.

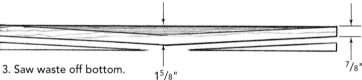

3. Saw waste off bottom.

4. Completed rail.

Tip: *If your mortises are not centered across the work, you can get frames that aren't flat. To avoid misalignment, mark the top face of each part. Orient that face against the mortising jig, and the mortises will be consistently located.*

Cutting the top rails and runners

1. If you didn't do this earlier, rip and crosscut the top rails to size. Bear in mind that the rails of the top frame are longer than all the other rails, because the top frame is dovetailed into the post tops.

2. Lay out and cut the large tails on the ends of the top drawer and back rails (see "Web Frame Construction" on p. 70). The cutting can be done with a jigsaw or on the bandsaw.

3. Rip and crosscut all the runners to the dimensions given in the cut list. Note that they are the same thickness as the rails.

Photo I: The V-shape of the bottom rail is routed into a thick workpiece. Secure the work and the template, which is screwed to a fence, in a vise for routing. The rail's bottom edge is then cut to shape on the bandsaw.

Cutting the joinery

The rails and runners are assembled into web frames held together with loose tenons (see "Web Frame Construction" on p. 70).

1. Lay out the mortise locations on the rails and runners. The top three rails (both front and back) have three mortises each. To accommodate the dovetails, the top rails—the long ones—have the end mortises located an extra ⅝ in. farther in from their ends. The runners should be identical to one another.

2. Set up the mortising jig and plunge router for mortising the rails. Clamp a rail to the work holder and adjust the edge guide and the jig's stops. Mark the extents of the mortises on the jig itself; then use those marks to position each rail and runner for mortising.

3. Rout the mortises in the rails.

4. Change from the horizontal work holder to the vertical one.

5. Rout the mortises in the runners.

Shaping the bottom rail

The bottom drawer rail has a V-shape to it, giving the face of the chest a pointy chin. The entire rail, of course, isn't shaped, only the underside. The upper surface has a recess cut into it to accommodate the bottom drawer. The rail should already have been cut to size and then mortised for joining to the runners and the posts. Now you must shape it.

1. Make the template for routing the recess (see "Making the Bottom Drawer Rail"). Mount the template to a fence, such as a straight 2x4.

2. Rout the recess. The template is designed to be held in place by clamping both it and the workpiece in a bench vise (see **photo I**). Use a pattern bit (which has a shank-mounted pilot bearing) to make the ¾-in.-deep cut.

3. Chisel the tight inside corner of the recess. Even a small-diameter bit won't produce a sharp corner, which is what you want. So sharpen the rounded corner left by the router bit by paring it with a chisel.

4. Cut the bottom surface parallel to the top edge of the recess. Mark the cuts on the edge of the rail and then saw to the lines on the bandsaw.

5. Plane or sand the cut surfaces smooth and free of saw marks.

Tip: The back rails are ¼ in. narrower than the front rails. To maintain a consistent depth for the mortises without resetting the router's plunge depth, use a ¼-in.-thick spacer under the back rails when clamping them in the mortising jig.

Assembling the frames

1. Make the loose tenons needed for the web frames and fit them to the mortises.

2. One by one, assemble the frames without glue. Apply clamps and measure both diagonals to ensure that each clamps up square. (If the diagonal measurements are equal, the assembly is square.) If it doesn't do that without glue, it isn't going to do it after the glue's been applied. Make whatever adjustments are necessary to ensure all the frames will be square after glue-up.

3. Glue and clamp the frames. Re-check the diagonal measurements, of course, and make sure the frames are flat.

4. After removing the clamps, scrape and sand the frames to remove any dried squeeze-out and smooth the top surface.

Cutting the drawer divider joinery

Vertical drawer dividers are typically joined to the rails with dadoes or sliding dovetails. In this chest, loose tenon joints do the job.

1. Lay out the mortises on the front rails of the top web frame and the two web frames with the center runner. Each mortise must be equidistant from the ends of the rail and perpendicular to the front edge of the rail it is cut into. The middle frame must be mortised twice, in the topside and underside of the rail, to support the dividers above and below it.

2. Set up the plunge router with the $\frac{5}{16}$-in. straight bit. Since one web frame is mortised in both the topside and the underside, the mortises can't be very deep. The cut depth should be no more than $\frac{3}{8}$ in.

3. For each cut, clamp a fence to the frame to guide the router (see **photo J**).

4. Rout the mortises.

5. Cut the vertical dividers to size.

6. Mortise the dividers using the mortising jig and plunge router setup.

Photo J: The vertical drawer dividers are joined to the web frames with mortises and loose tenons. Rout the mortises after the frames are glued up.

ASSEMBLING THE CASE

With the major subassemblies completed, it is just about time for the case to come together. You need to make all the loose tenons that link the web frames and the side. Then comes a dry fitting, followed by the final glue-up.

Making the loose tenons

The sides and web frames are linked with forty-eight 1½-in.-long loose tenons.

1. Select strips of stock and mill them down to ⁵⁄₁₆ in. thick.

2. Rip the strips to ½ in. wide.

3. Nose the edges with a ⅛-in.-radius roundover bit, so they'll fit the mortises.

4. Crosscut the strips into 1½-in.-long loose tenons.

Joining the sides and web frame

The case, by virtue of all its parts, is a complex assembly. Putting the parts together without glue, and clamping the assembly, is an important prelude to the final glue-up. Dry-fit the case and clamp it tight, just as though you had used glue. Satisfied that everything fits properly and that you've got the assembly routine down, dismantle the chest and reassemble it, but this time use the glue.

1. Set one side assembly on the benchtop, mortises up.

2. Slip a loose tenon in each mortise in the side assembly.

3. One by one, stand all the web frames in their proper positions. The bottom frame should be at the foot end. Set the vertical dividers in place too.

4. Set the second side assembly atop the web frames and work the tenons into its mortises (see **photo K**).

5. Stand the assembly upright on the floor. The tenons should be tight enough in the mortises that the assembly stays together. (If the tenons are too loose, make a new batch.)

6. With a spring clamp, secure a long ripping to the outer face of each post. Apply bar or pipe clamps across the case, aligned with each drawer rail and back rail. Tighten them, just as you will when assembling the unit with glue (see **photo L**).

7. Check over the case. Use a square and diagonal measuring to make sure it is case is square. Be sure all the rails seat tight to the posts.

8. Dismantle the case; then reassemble it using glue.

Photo K: Case assembly begins with the side flat on the bench. After joining all the web frames to it, set the second side in place and methodically start all the tenons into their mortises.

Photo L: After the second side is seated, right the unit and apply clamps parallel to each web frame, across the front and the back. The spring clamps hold pad strips that protect the posts.

Photo M: A large dovetail cut on each end of the top drawer and back rails join them to the posts. This traditional joinery captures the rails in the posts.

Dovetailing the top frame

The top web frame is dovetailed into the post tops. The tails have already been cut. Now you'll scribe the locations onto the tops of the posts and, with the case clamped together, cut the dovetail slots.

1. Set the top web frame on top of the posts. Make sure it is properly aligned.

2. Scribe around the tails with a marking or utility knife. Make sure you will be able to see the marks. Measure and mark the depth of slot necessary, and extend the knife marks down the inner surface of the post.

3. Chisel out the waste.

4. Carefully pare the slots until the frame can be seated. It should be flush with the tops of the posts (see **photo M**).

Making the top

The top is a panel formed by edge-gluing several boards. The front and ends are beveled, to make it appear thinner.

1. If necessary, glue up boards to form the top panel.

2. Scrape and sand the panel to remove excess glue and to smooth and flatten it.

3. Rip and crosscut the panel to final size

4. Lay out the bevels on the front and one end. Note that they are of different widths.

5. Cut the bevels. Do this on the table saw by tilting the blade and then standing the top on edge and feeding it along the rip fence.

6. Scrape and sand the bevels to smooth them and remove saw marks.

7. Mount the top with screws driven up through the top web frame into the underside. The pilot holes you drill through the back rail should be sized for the screws; this will fix the top at the back. In the runners and the front rail, the pilots should be oversize or oblong, so the panel can expand and contract.

Making the back

The next logical step is to cut and install the back panel. But you may prefer to have access to the back of the case while you make and fit the drawers. You can cut the panel and install it now or install it later.

1. Measure the case from rabbet to rabbet, and from bottom web frame to top.

2. Cut a piece of ¼-in. birch plywood to fit.

3. Mount the back panel by driving three screws through it into each back rail.

BUILDING THE DRAWERS

The final phase of the project is building the drawers. All are constructed in the same way (see "Drawer Construction"). Although there are eight drawers in the chest, there are only four different sizes.

Cutting the parts

The drawers are built of the usual front, back, sides, and bottom. But these drawers also have slips, which are glued to the drawer sides to broaden the bearing surface and reduce wear on the sides and the runners.

The fronts are made of cherry, the bottoms of ¼-in. birch plywood, and the remaining parts of a secondary wood. Burton used soft maple as his secondary wood.

DRAWER CONSTRUCTION

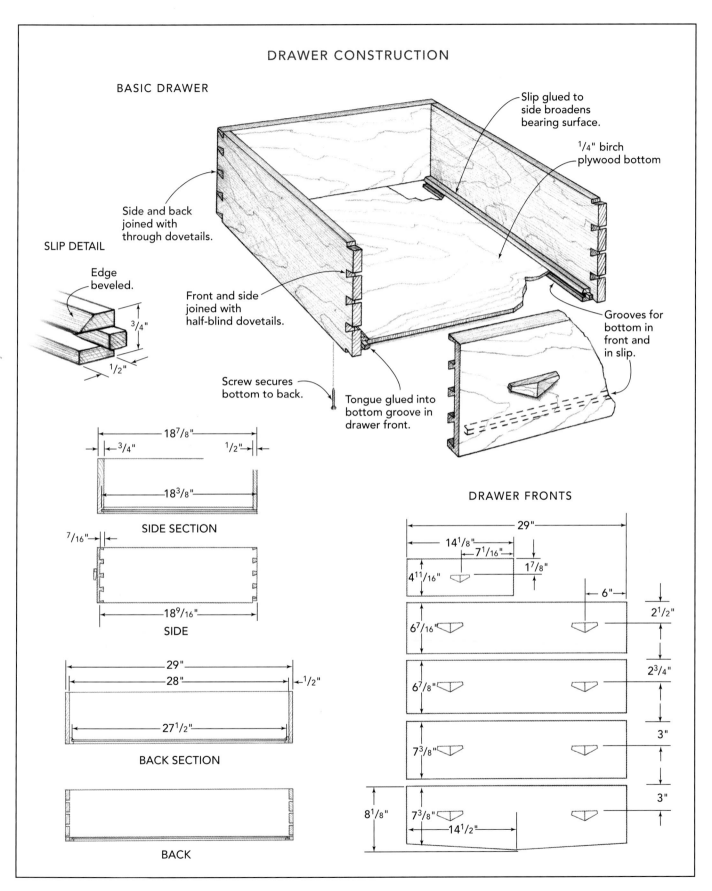

BASIC DRAWER

Slip glued to side broadens bearing surface.

1/4" birch plywood bottom

Side and back joined with through dovetails.

SLIP DETAIL

Edge beveled.

3/4"

1/2"

Front and side joined with half-blind dovetails.

Screw secures bottom to back.

Tongue glued into bottom groove in drawer front.

Grooves for bottom in front and in slip.

18 7/8"

3/4"

1/2"

18 3/8"

SIDE SECTION

7/16"

18 9/16"

SIDE

29"

28"

1/2"

27 1/2"

BACK SECTION

BACK

DRAWER FRONTS

29"

14 1/8"

7 1/16"

1 7/8"

4 11/16"

6"

2 1/2"

6 7/16"

2 3/4"

6 7/8"

3"

7 3/8"

3"

8 1/8"

7 3/8"

14 1/2"

1. Face joint and thickness plane stock for the drawer fronts, sides, backs, and slips. The fronts and slips are ¾ in. thick, while the sides and backs are ½ in. thick.

2. Rip and crosscut the drawer fronts. Before turning on the saw, compare the dimensions specified on the cut list with the actual dimensions of the drawer openings of your chest. As you cut the fronts, fit them to their openings. Side to side, the fronts can be closely fitted. But you need to allow room for the front to expand across its grain; and the taller the drawer opening, the greater the allowance.

3. Rip and crosscut the sides and backs. As with the fronts, reconcile the dimensions specified on the cut list with the actual dimensions of your chest before you cut anything. Cut the parts to fit the openings and label them.

Cutting the joinery

Several different joinery cuts are required. The fronts and sides are joined with half-blind dovetails. The sides and back are joined with through dovetails. The bottoms ride in grooves in the slips and drawer front. As usual, there are a variety of ways you can make the cuts.

1. Cut a groove for the bottom in the front and the slips. Measure the plywood's actual thickness and try to match it when cutting the grooves. Plow the grooves end to end.

2. Traditionally, the front end of each slip is rabbeted across the top and bottom surfaces to produce a short tongue that fits into the panel groove in the drawer front (see "Drawer Construction" on p. 83). Similarly, the tail end is rabbeted across the top surface to accommodate the drawer back and allow the slip to end flush with the back end of the drawer side. Neither cut is essential, of course. But if you make the slips in this way, it's fairly efficient to cut the tongue on the front; then crosscut the slip to fit when its drawer is dry-assembled. Then you can rabbet the end.

3. Before setting the slips aside, cut a 45-degree chamfer on what will be in upper inside edge.

4. Lay out and cut the half-blind dovetails that join the fronts and sides. The dovetails can be

Photo N: Burton saws the margins of the pins; then routs out the majority of the waste. A strip of plywood with a glued-on fence controls the front-to-back depth of cut, and the side-to-side extent of the cut is done freehand.

Photo O: The final paring is done with a pair of custom-ground chisels: one for the left, one for the right.

Photo P: The pins on the drawer backs are routed using Burton's shopmade adjustable pin-cutting template.

hand cut or machined with a router and dovetail jig. Burton uses a router to excavate the bulk of the waste quickly from the recesses in the fronts (see **photo N**). He then pares them to fit with a pair of old chisels ground with the cutting edges angled to the left and right (see **photo O**).

5. Lay out and cut the through dovetails that join the sides and backs. Bear in mind that the drawer bottoms slide just beneath the backs, so the bottom half-pin of the back penetrates the side ½ in. above the bottom edge. Again, use your favorite approach to cutting these joints. As he does in cutting half-blind dovetails, Burton combines the use of a router and shopmade guide (see **photo P**) with hand fitting.

Photo Q: Another traditional element in Burton's chest is the use of slips to broaden the bottom edges of the drawer sides. The extra bearing surface will greatly improve the life of the drawers and the runners.

Assembling the drawers

Each drawer is assembled according to the same routine. Take the time to get each one square and to fit it to its opening.

1. Run through a dry-assembly before actually gluing up each drawer.
2. Crosscut each slip to length and then rabbet its end.
3. Next, glue the slip to the inside face of the drawer side, flush with the bottom edge (see **photo Q**).

4. Glue up the drawer.
5. Measure each drawer for a bottom. Cut the bottoms.
6. Slide the bottoms into the drawers. Secure them by driving a screw through the bottom into the center of the back.
7. Fit each drawer to its opening. Sand or plane the sides as necessary to get the drawer operating smoothly (see **photo R**).

Photo R: With the pins deliberately cut a hair short, the sides are just proud of the pin ends in the assembled drawer. Burton can quickly handplane them flush with the pins.

FINISHING UP

Burton designed his own cherry pulls for this chest and mounted one in the center of each drawer front. You could, of course, use commercially made pulls (see "Sources" on p. 216). If you are using non-wooden pulls, remove them before finishing the chest.

Everyone has a favorite finish. Burton's is oil. It is a great choice for the small-shop woodworker. It doesn't require specialized equipment to apply. Burton finished his chest inside and out with several coats of the oil. The surfaces of the drawer fronts—meaning the face and the edges—are oiled, while the rest of each drawer is shellacked.

BOW-FRONT CHEST

Such a modest-looking chest, but such an exercise in woodworking. It really is deceptive. This mahogany chest has a gentle bow to the front. The size is relatively small, with only five drawers. Yes, it has a bit of inlay, but it seemed a fairly quick-and-easy bow-front chest to fill out my project lineup. Then I talked to the craftsman, Mark Edmundson.

It turns out that this project has it all: sawing veneer, applying veneer, inlay, bent laminations, hand-cut dovetails on curved pieces, out-of-the-ordinary drawer construction, mortises and loose tenons, splined joints, and router template work. If I tell you that it was Edmundson's final project as a second-year student at the College of the Redwoods, perhaps you'll understand. It was intended to prove what he had learned.

Influenced by the work of Arts and Crafts designers Edward and Sidney Barnsley, Edmundson followed their use of inlay to break up broad surfaces, and he crafted pulls reminiscent of a bronze drop pull that they used. Those are subtle influences, but the entire design is subtle. The proportions are subtle. Just as the drawer heights are scaled, the thicknesses of the drawer dividers are scaled. And the placement of the drawer-front inlay is scaled. It's another aspect of the deception. You don't pick this proportioning out, but your eye appreciates it nonetheless.

Bow-Front Chest

WITH LOOSE TENONS, biscuits, and splines in the primary case joinery, the bow-front chest is untraditional but nonetheless uncomplicated to build. The side assemblies—posts and rails and veneered panels—are joined with a bottom assembly of rails and a plywood bottom. The top web frame is dovetailed into the posts, whereas the others are slid into place on splines. The NK-style drawers are unusual and a challenge to build. What elevates this project is the veneering, the inlay, and the bent laminations.

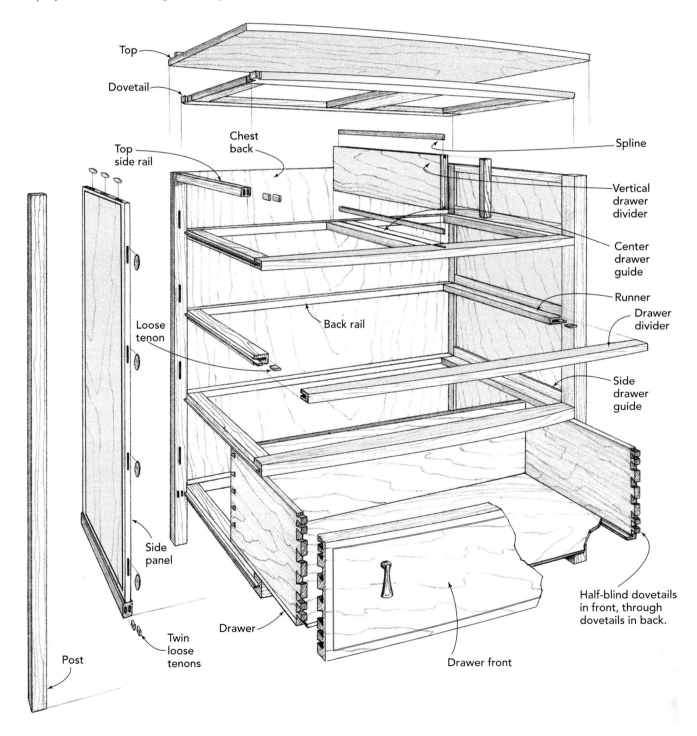

Top

Dovetail

Top side rail

Chest back

Spline

Vertical drawer divider

Center drawer guide

Runner

Drawer divider

Back rail

Loose tenon

Side drawer guide

Side panel

Half-blind dovetails in front, through dovetails in back.

Post

Twin loose tenons

Drawer

Drawer front

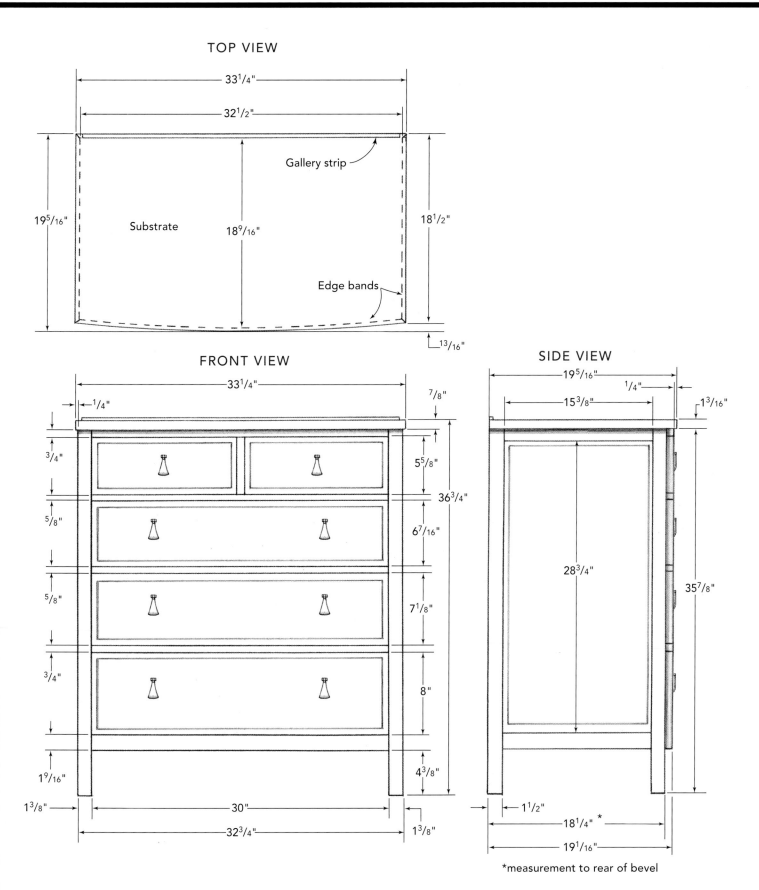

TOP VIEW

33^1/$_4$"

32^1/$_2$"

Gallery strip

19^5/$_{16}$"

Substrate

18^9/$_{16}$"

18^1/$_2$"

Edge bands

^{13}/$_{16}$"

FRONT VIEW

33^1/$_4$"

^{7}/$_8$"

^{1}/$_4$"

^{3}/$_4$"

5^5/$_8$"

36^3/$_4$"

^{5}/$_8$"

6^7/$_{16}$"

^{5}/$_8$"

7^1/$_8$"

^{3}/$_4$"

8"

1^9/$_{16}$"

4^3/$_8$"

1^3/$_8$"

30"

1^3/$_8$"

32^3/$_4$"

SIDE VIEW

19^5/$_{16}$"

^{1}/$_4$"

15^3/$_8$"

1^3/$_{16}$"

28^3/$_4$"

35^7/$_8$"

1^1/$_2$"

18^1/$_4$" *

19^1/$_{16}$"

*measurement to rear of bevel

BUILDING THE CHEST STEP-BY-STEP

CUT LIST FOR BOW-FRONT CHEST

Case			
4	Leg posts	1⅜ in. x 1½ in. x 35⅞ in.	mahogany
2	Bottom side rails	1¼ in. x 1⁹⁄₁₆ in. x 15⅜ in.	mahogany
2	Top side rails	1¼ in. x 1³⁄₁₆ in. x 15⅜ in.	mahogany
1	Bottom front rail	1⁹⁄₁₆ in. x 2¹⁄₁₆ in. x 30 in.	mahogany
1	Bottom back rail	¾ in. x 1⁹⁄₁₆ in. x 30 in.	mahogany
4	Loose tenons	¼ in. x 1¹⁄₁₆ in. x ¾ in.	soft maple
4	Loose tenons	¼ in. x ¹¹⁄₁₆ in. x ¾ in.	soft maple
4	Loose tenons	½ in. x 1¹⁄₁₆ in. x 2 in.	soft maple
16	Loose tenons	¼ in. x ¾ in. x 2 in.	soft maple
4	Loose tenons	¼ in. x 2½ in. x 2 in.	soft maple
1	Bottom panel	¾ in. x 15⅜ in. x 30 in.	birch plywood
2	Side panels	¾ in. x 15⅜ in. x 28¾ in.	birch plywood
1	Top	¾ in. x 18⁹⁄₁₆ in. x 32½ in.	plywood
1	Back panel	¼ in. x 30¾ in. x 30⁵⁄₁₆ in.	plywood
3	Edge band plies	⅛ in. x ¹⁵⁄₁₆ in. x 35 in.	mahogany
2	Edge bands	⅜ in. x ¹⁵⁄₁₆ in. x 18½ in.	mahogany
1	Edge band	⅜ in. x ¹⁵⁄₁₆ in. x 33¼ in.	mahogany
1	Gallery strip	¼ in. x ⅜ in. x 32 in.	mahogany
8	Drawer guides	⅜ in. x 1½ in. x 15⅜ in.	soft maple
1	Top drawer dividers	¾ in. x 2¹⁄₁₆ in. x 31¼ in.	mahogany
1	Top back rail	¾ in. x 1 in. x 31¼ in.	soft maple
2	Second/third drawer dividers	⅝ in. x 2¹⁄₁₆ in. x 30 in.	mahogany
2	Second/third back rails	⅝ in. x 2¹⁄₁₆ in. x 30 in.	soft maple
1	Fourth drawer divider	¾ in. x 2¹⁄₁₆ in. x 30 in.	mahogany
1	Fourth back rail	¾ in. x 1 in. x 30 in.	soft maple
4	Drawer runners	⅝ in. x 1¼ in. x 15½ in.	soft maple
4	Drawer runners	¾ in. x 1¼ in. x 15½ in.	soft maple
1	Center drawer runner	¾ in. x 3 in. x 15½ in.	soft maple
1	Center drawer runner	⅝ in. x 3 in. x 15½ in.	soft maple

CUT LIST FOR BOW-FRONT CHEST

6	Splines	¼ in. x ¾ in. x 17½ in.	plywood
2	Splines	¼ in. x ½ in. x 17½ in.	plywood
1	Vertical drawer divider	¾ in. x 5⅝ in. x 17¹³⁄₁₆ in.	plywood
1	Edge band	¾ in. x ¾ in. x 5⅝ in.	mahogany

Drawers

3	Bottom panels	¼ in. x 18³⁄₁₆ in. x 29⅜ in.	plywood
2	Bottom panels	¼ in. x 18³⁄₁₆ in. x 14 in.	plywood
8	Slides	½ in. x 1 in. x 17⁵⁄₁₆ in.	soft maple
2	Slides	½ in. x 1 in. x 17¹⁵⁄₁₆ in.	soft maple
1	Drawer front	⅝ in. x 8 in. x 30 in.	mahogany
1	Drawer front	⅝ in. x 7⅛ in. x 30 in.	mahogany
1	Drawer front	⅝ in. x 6⁷⁄₁₆ in. x 30 in.	mahogany
2	Drawer fronts	⅝ in. x 5⅝ in. x 14⅝ in.	mahogany
2	Drawer sides	½ in. x 7½ in. x 17¹¹⁄₁₆ in.	soft maple
2	Drawer sides	½ in. x 6⅝ in. x 17¹¹⁄₁₆ in.	soft maple
2	Drawer sides	½ in. x 5¹⁵⁄₁₆ in. x 17¹¹⁄₁₆ in.	soft maple
2	Drawer sides	½ in. x 5⅛ in. x 17¹¹⁄₁₆ in.	soft maple
2	Drawer sides	½ in. x 5⅛ in. x 18⁷⁄₁₆ in.	soft maple
1	Drawer back	½ in. x 7½ in. x 29⅞ in.	soft maple
1	Drawer back	½ in. x 6⅝ in. x 29⅞ in.	soft maple
1	Drawer back	½ in. x 5¹⁵⁄₁₆ in. x 29⅞ in.	soft maple
2	Drawer backs	½ in. x 5⅛ in. x 14½ in.	soft maple

Miscellaneous

8	Drop-type drawer pulls		
	Veneer	25 sq. ft.	mahogany
	Veneer	15 sq. ft.	soft maple
	Inlay	¹⁄₁₆ in. x ¼ in. x 42 ft.	maple, alder, or holly

Drawer fronts are glue-laminated from veneers. Calculate materials needed based on the given dimensions and the thickness of veneer used.

SUBTLE AS IT IS, the bow of this chest's front dictates the construction sequence. You need to create the drawer fronts before you build anything, so you can determine what the exact contour of the bow will be. The drawer fronts are glue-laminated over a curved form, and they will spring back slightly from the contour of the form. Until you have at least one drawer front laminated and off the form, you won't know exactly what the bow contour will be.

Photo A: Use a slow, steady feed rate when sawing the stock. Keep the work tight to the fence with pressure applied just ahead of the cut. Using a push block is an especially good idea when the workpiece gets thin (down to the final two or three veneers).

The drawer fronts are made with a bent lamination process, which isn't difficult but does require you to make a very sturdy form that will resist the bending forces. The drawer laminates themselves can be resawn from wood that's a close match to the face veneer (see **photo A**). If you like, you can also make the drawer fronts entirely from commercially available veneer, using a less-expensive grade for the inner plies and choice veneer for the exposed surface. You can also do what Edmundson prefers and resaw the drawer fronts and all the veneer from the same stock.

The panels and top for the chest are veneered as well, but if you don't have a veneer press or a vacuum press, you can still build the chest by substituting choice mahogany plywood for the panels. The downside here is that the panels and drawer fronts may not match as exactly as the original. But it will still be a beautiful chest.

MAKING THE DRAWER FRONTS

Each drawer front is formed of several layers (or plies) of the veneer. Glue is spread on the plies. They are stacked up and then bent over a form and clamped there until the glue cures.

One of two forming approaches can be used. The more common employs two forms: a convex one and a concave one. The plies to be laminated are placed between these forms, and then the forms are clamped together, forcing the plies into the curved contour. The alternative approach requires only the convex form, and it uses a vacuum bag to force the lamination to the form and hold it there while the glue cures.

I'm going to outline a good way to make the two forms. If you have a vacuum bag and want to use it, then you need to make only the convex form and can skip over some of the steps that follow.

The laminating process itself doesn't take a great deal of labor time, but it *does* take hours for the glue to set. Edmundson basically did one drawer front a day. Once the first front was off his form, he could capture its curve and thus could move ahead with chest construction (while still doing the other drawer fronts).

Making the drawer-front template

Construction of the drawer fronts begins with a curve, the curve you *want* the front of the chest to have. Edmundson sketched an elliptical curve as his starting point. You may opt for more of a fixed-radius curve, if only because it's easier to lay out in the first place. You then make a template of the curve. That template is the one you use when shaping the drawer dividers and making the additional pair of templates for shaping the laminated plies.

1. Cut a template blank 6 in. to 7 in. wide and 35 in. long from ¼-in. plywood or hardboard.
2. Lay out the crest and the end points of the curve on the blank; then lay out the curve. The crest is at the center of the blank, of course, and the end points are ¹⁵⁄₁₆ in. back at each end of the blank. An easy way to produce a curve is to flex a thin ripping so it is aligned with each of the three marks. Have a helper trace along the curved strip onto the template blank.
3. Cut to the line on the bandsaw or with a jigsaw. Smooth the curve with a spokeshave, file, or coarse sandpaper. Label this template as the master.

Making the form templates

To make the laminating forms for the drawer fronts, you need two templates, a concave one and a convex one. The concave drawer front template must have the exact curve of your master template; in other words, it must be the positive of the negative master. The convex master template must be a curve that complements the first, but accounts for the ⅞-in. drawer front thickness.

"Making the Laminating Forms" outlines the steps you need to follow when making the templates using the router (see **photo B** on p. 96). This method will yield very accurate templates.

On the other hand, it is surely quicker to simply trace the master template's arc onto a wide piece of plywood, split the line on the

MAKING THE LAMINATING FORMS

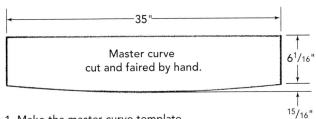

Step 1. Make the master curve template.

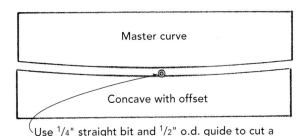

Step 2. Use master curve to make concave offset template.

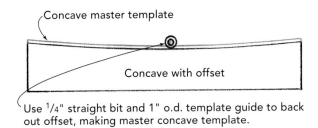

Step 3. Back out the offset, making a concave master.

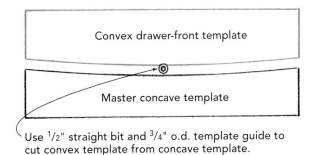

Step 4. Make convex drawer front template from concave master.

Photo C: Assembling the laminating form is a matter of applying glue to the face of a ply, sliding it onto the registration dowel, squaring it, and then driving two or three screws through it into the previous ply.

Photo B: Making a convex template to complement the concave one is quick and easy. Use your router and template guides (or use a pattern bit of the correct diameter).

bandsaw or with a jigsaw, and use the resulting two pieces as the templates for making the forms. Templates made this way won't be as precise, but the fact is, they'll work satisfactorily.

Constructing the laminating forms

Each part of the form is constructed of 12 plies of ¾-in. medium-density fiberboard (MDF) or particleboard. The plies are cut to shape using the templates. Then they are glued together and sanded smooth.

1. Cut 24 pieces of ¾-in. MDF or particleboard to 4 in. by 33 in. You'll use 12 pieces in each half of the form.

2. Set up the drill press with a fence and stop so you can drill a ¾-in.-diameter registration hole through each piece. Of course, the holes must be in exactly the same spot on each. Drill a matching registration hole through the templates, too.

3. Cut the template plies. Use a ¾-in.-diameter dowel to register the template on an MDF blank. Clamp the two together. With a jigsaw or bandsaw, cut the MDF close to the template. Then use a router and a straight pattern bit, which has a shank-mounted pilot bearing, to rout the MDF to match the template's curve. Cut all 24 pieces needed for the two forms in this way.

4. Assemble each form. Glue the MDF plies together. Use the dowel to register them as you glue them. As you build up the form, ply by ply, run two or three screws through each ply into the previous one (see **photo C**).

5. Sand the curved surface of each form to smooth any ridges or high spots and to blend all the plies into a continuous fair curve.

6. Wax the forms. This will make it easier to clean glue off the forms and will help prevent the lamination from bonding to the form.

Laminating the drawer fronts

The drawer fronts that Edmundson made are about ⅝ in. thick. He uses plies cut to the same thickness as his face veneer—approximately ¹⁄₁₆ in. Nine plies of this veneer will glue up to a thickness of just about ⅝ in. Other woodworkers use thicker plies for bent laminations, and you would need fewer of those to get the ⅝ in. thickness.

To give the appearance of a solid board in the finished lamination, glue the veneers together in the order that they came off the saw. This helps hide glue lines and gives a more natural look to the work.

Note: It's essential to work quickly during a glue-up. You must have the forms at hand, well waxed, and ready for the laminate bundle. Your clamps also must be ready. I set the lamination form on a corner of my workbench and an adjustable workstand, so I'd have unob-

structed access to the edges to apply the clamps (see **photo D**).

1. Lay out the veneers to be joined in the lamination. Organize them however you want to arrange them in the glue-up, and work out a routine for keeping them in the proper order and proper orientation as you spread the glue.
2. Apply glue to the mating surfaces of the veneers. That means both sides of all but the two face veneers, which are coated on only one side. I spread glue on the first face with the veneer resting on the benchtop. The very first piece goes on the form with the glue-side up. All subsequent ones are set onto the stack on the form, glue-side down. Then spread glue on the second face. Cover the surfaces completely, but don't get carried away. You need to expect beads of glue to squeeze out of the seams as you apply the clamps. But rivers of

Tip: You can make a small amount of showy primary wood go much further if you make the drawer fronts with inner plies of a secondary wood and use veneers sawed from the primary wood for the face only.

Photo D: Spreading glue on the veneers doesn't have to be a messy job. Apply glue to one face of the veneer, then turn it onto the stack and apply glue to the second face. The veneers may squirm a bit, but the tack of yellow glue will prevent wholesale slippin' and slidin'.

Photo E: Clamping the two forms to bend the veneers isn't much different from clamping a panel. Begin by tightening the center clamp. Tighten the clamps on either side next, followed by those at the very ends.

glue welling from the seams indicate you used way too much glue. It's a nightmare to clean up. A 3-in. low-nap paint roller is a good tool for spreading the glue (if you can find what amounts to a no-nap roller, use that).

3. Set the convex form on top of the stack and apply the clamps. Work as quickly as possible. You might picture the plies slippin' and slidin' and the wood resisting being bent, but that's not the case here. If you use yellow glue, its tack will keep the individual plies from slithering far out of alignment. The curve is gentle enough that there isn't much resistance to the bending.

Apply bar clamps, starting at the center and working toward the ends (see **photo E**). The amount of movement needed to pull the laminates to the form is small, well within the range of even pipe clamps.

4. Allow time for the glue to set fully. Plan on leaving the drawer front in the form 8 hours to 12 hours. If you do one front a day, you can give the glue nearly 24 hours to set.

5. Free the drawer front from the forms. Remove the clamps, separate the lamination from the form, and set it aside until you are ready to make the drawers. Clean any dried glue from the forms, rewax them, and do another lamination.

Making the bow template

With the first drawer front off the form, you can capture its curve and make a template to use in making the chest's case.

1. Joint one edge of the drawer front so it will set flat on your template stock. Do this with the convex face against the jointer fence, moving the work so the face stays in contact with the fence as it is advanced across the knives. But bear in mind here that you are after a straight edge, not necessarily one that's square to the face of the drawer.
2. Stand the lamination on a piece of ¼-in. plywood. Trace along the edge.
3. Cut to the line on the bandsaw. Then smooth the edge.
4. Trim the template to a size commensurate with its use. You won't rout the curve on the drawer rails until after the web frames are assembled, so you want to leave the template pretty big. That way you can clamp it to a web frame and not have to worry about the clamps interfering with the router.

CONSTRUCTING THE CASE

The case is a modern sort of post-and-panel construction. The posts and rails are assembled with mortise-and-loose-tenon joints. The side panels are plywood joined to the post-and-rail assemblies with biscuits. And to make the construction work interesting, there are the veneering and inlay operations and the fitting of the bow-front web frames to the case. Construction begins with the basic parts: the posts and rails.

Cutting the parts

The posts, the side rails, and the front bottom rail are all cut from 8/4 mahogany. The back bottom rail is cut from 4/4 mahogany.

1. Rip and crosscut the parts to rough sizes, as appropriate, before dressing the stock.
2. Joint a face and an edge of each piece.
3. Plane the stock to the correct thickness as indicated in the cut list. The posts are all

1⅜ in. by 1½ in. If you cut individual blanks from the rough stock, you can make them the final width and thickness using the thickness planer. Note that the side rails, front bottom rail, and back bottom rail are different thicknesses.
4. Rip the pieces to the widths specified on the cut list.
5. Bevel the front posts. These posts must be shaped to match the bow of the case front. Since the curve is really very shallow, Edmundson simply beveled the front edges of the posts on the table saw. Tilt the blade to 5 degrees and rip the front two posts.

Routing the post-and-rail joinery

All the post-and-rail joints use loose, or slip, tenons. To make the joinery, you cut mortises in both mating pieces and then join the pieces with a separate tenon that is glued into the mortises. Since you will probably use the same tool to do all the mortises, do them together. Edmundson uses a horizontal boring machine, but a plunge router also works well for the repetitive mortising task. See "Case Side Joinery" on p. 100 for dimensions and placement of these joints.

1. Cut the ½-in.-wide by 1¹⁄₁₆-in.-long by 1-in.-deep mortise in the post for the bottom front rail.
2. Cut the matching mortises into both ends of the front rail. The length of the rail makes the staging a bit of a challenge if you do this task with a router.
3. Change your tool setup to cut ¼-in.-wide mortises next.
4. Begin by cutting the deep mortise in the back posts for the bottom rail.
5. Cut matching mortises in the rail ends.
6. Cut the twin mortises next. Positioning the mortises consistently is critical. With my router setup (see "Mortising Jig" on p. 124), I do all the near mortises first; then all the far mortises. *Near* and *far* refer to the mortise's position relative to the reference face of the workpieces.

CASE SIDE JOINERY

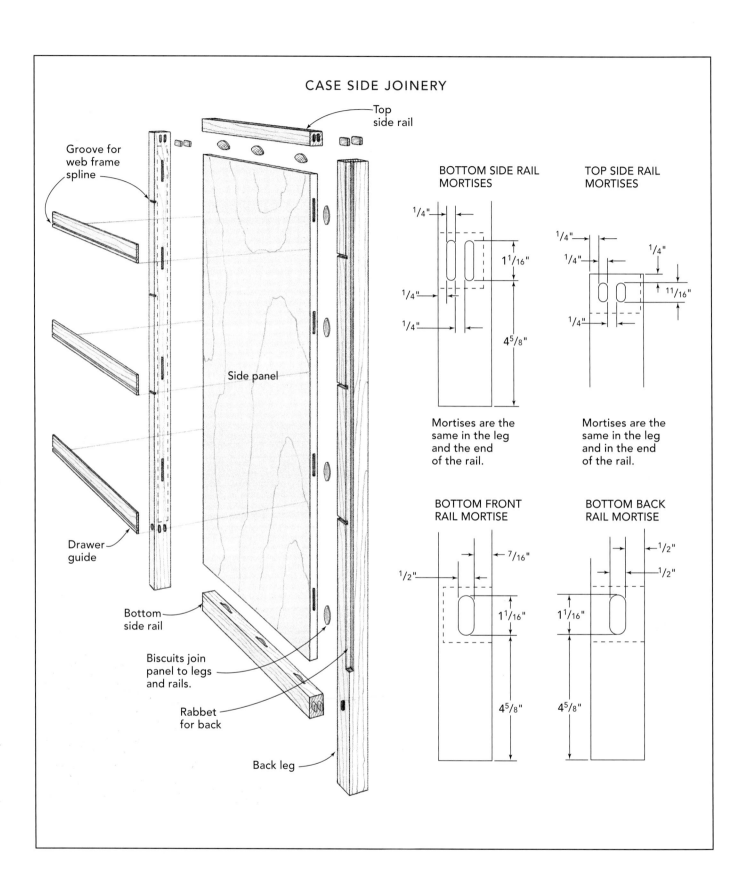

Top side rail

Groove for web frame spline

Side panel

Drawer guide

Bottom side rail

Biscuits join panel to legs and rails.

Rabbet for back

Back leg

BOTTOM SIDE RAIL MORTISES

1/4"

1 1/16"

1/4"

1/4"

4 5/8"

Mortises are the same in the leg and the end of the rail.

TOP SIDE RAIL MORTISES

1/4"

1/4"

1/4"

1/4"

11/16"

Mortises are the same in the leg and in the end of the rail.

BOTTOM FRONT RAIL MORTISE

7/16"

1/2"

1 1/16"

4 5/8"

BOTTOM BACK RAIL MORTISE

1/2"

1/2"

1 1/16"

4 5/8"

Making the loose tenons

Make a list of the mortises you've just cut and note their dimensions. Make the loose tenons necessary to assemble the posts and nails.

1. Mill scraps of soft maple to the final thicknesses, using pieces that are long enough to be machined safely. Be sure the material is wide enough for the applications, of course. The front rail mortises require ½-in.-thick stock. All the others require ¼-in. stock.
2. Rip the stock to the needed widths.
3. Round the edges of the tenon stock on the table-mounted router. Use either bullnose or roundover bits.
4. Crosscut the tenons to length.

Dry-assembling the legs and rails

Even if you can't assemble all the frame parts at once, you can join selected ones into subassemblies. The point of the exercise is to make sure everything fits, of course, and that the resulting assemblies are square and flat. But you also want to measure for the panels. (You don't want to glue up the bottom panel and rails, then discover that the mortises don't line up.)

1. Slip the tenons into the mortises in the side rails.
2. Fit the legs onto the tenons.
3. Clamp the assembly. If you have metal-jawed clamps that might mar the work, use cauls between jaws and wood.
4. Check the assembly for squareness and flatness.
5. Measure the panel space carefully and note the dimensions.
6. Assemble the second side in the same way. Note the dimensions of its panel space.
7. Insert the tenons into the front and back rails.
8. Join the rails to the two clamped-up side assemblies. Apply clamps, paralleling the rails.
9. Check the assembly for squareness and flatness.
10. Measure the space for the bottom panel. Note the dimensions.

Cutting the panels

The chest actually has quite a few panels. You think first of the side panels and the top. But it also has a bottom panel, a back and even drawer bottoms. All are cut from plywood, and Edmundson veneered all the panels (but the bottom), which is why they come up here.

You need to cut, veneer, and inlay the side panels now so you can complete the case. While you have the veneering equipment out, you might as well veneer the top and—if you intend to veneer them at all—the back and the drawer bottoms.

1. Cut the bottom panel from ¾-in. plywood. Baltic birch plywood is ideal, but any good birch plywood is satisfactory. Use the dimensions you took from the dry-assembled case framework.
2. Cut the two side panels. Because you will veneer these panels, you should cut them about 1 in. wider and longer than the dimensions taken during dry assembly. Cut them to final size after they are veneered and before you do the inlay.
3. Cut the top. Although front edge of this piece is curved to match the drawer fronts and drawer dividers, cut it square and oversize by 1 in. or 2 in. until the veneer has been applied to the faces. Then you can trim it slightly undersize, shape the front edge, and apply the edge bands.
4. Cut the back panel and the drawer bottoms. If you are opting to veneer any of these parts, leave them 1 in. longer and wider than the cut list dimensions.

Veneering the side panels

The side panels are broad enough that you will need two or more veneer leaves per face. As you begin the veneering process, examine your veneers carefully to assess what you have and to determine the best combinations. If you can cover each panel with two leaves, you should consider book-matching. Slip-matching and reverse-matching are other options.

After you've identified the particular leaves you will use and how they'll be oriented, you can veneer the panels.

Photo F: The pieces of veneer must be joined edge to edge before they are glued to the substrate. Apply glue to only the substrate (not the veneer). Carefully move the veneer onto the substrate, align it, and tape it at the top and bottom edges.

1. Joint the edges of the leaves. There are all sorts of methods for doing this, but the most straightforward is to make a shooting board with a stop across one end. Rest the mating veneers on the shooting board, one with its face side up, the other with the face side down. With the jointer plane on its side on the benchtop, slide it along the shooting board, jointing the veneers.

2. Join the face veneers into a single sheet to cover the panel. A straightforward approach here is to edge-glue the pieces of veneer, clamping them with a simple system of straightedges and wedges. If necessary, you can prevent the panel from bowing upward by stretching some waxed paper over the seam and then weighting it. A metal plane works well here.

3. Joint and assemble the backing veneers into a single sheet.

4. Apply glue to one side of the substrate. (Spread the glue on only the substrate. If you spread glue on the veneer, it'll promptly curl up.) Use a low-nap paint roller or a notched spreader.

5. Lay the appropriate veneer in place (see **photo F**). If you have used veneer tape to hold individual pieces of veneer together, be sure the tape side is up. Secure the veneer to the panel with a piece of masking tape in the center at each end. This keeps the sheet from shifting as clamping pressure is applied, but it will be trimmed away when you size the panel.

6. Turn the panel over and repeat the process to apply the backing veneer.

7. Slide the panel into a vacuum press (see **photo G**) or a caul-and-platten veneer press and keep the panel under pressure for 4 hours or more, depending on the glue you've used. "Overnight" is the typical schedule.

8. Trim the veneered panel to its final dimensions. It's a good idea to make the cross-grain cut or cuts first and then the cut along the grain last. This should remove the minor tearout that occurs at the end of the cross-grain cut.

9. If you are opting to veneer the case back and/or the drawer bottoms, do it now, while you have the equipment out.

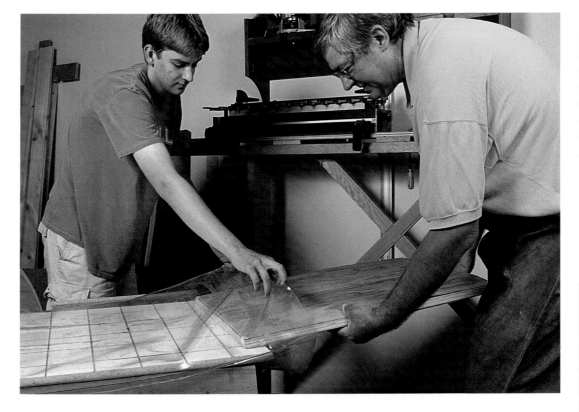

Veneering the top

The faces of the top are veneered in the same manner as the side panels. After the faces are veneered, you trim the panel to size, shape the front edge, and then apply the edge bands and gallery strip. Having the edge bands overlay the edges of the veneers provides the best protection for them.

1. Veneer both faces of the substrate, with the show veneers on one side and the secondary-wood veneers on what will be the bottom side.

2. Trim the veneered panel to the dimensions specified in the cut list.

3. Shape the front edge of the top. Using the master template, lay out the arc on the front edge of the panel. Do this with a pencil. Cut close to the line; then use the template and a flush-trimming bit to establish the final contour.

4. Using the concave master template for the drawer front laminating form, cut a clamping caul to use for banding the curved front edge of the top. Use ¾-in. MDF for the caul, and make it a couple of inches wide. Wax it thoroughly.

5. Cut the edge bands. To band the curved front edge, Edmundson used three ⅛-in.-thick plies of veneer, much like a bent lamination. Cut strips of the veneer, making them about 1 in. wide and a couple of inches longer than the surface to be banded. For the sides and back, cut strips of the primary stock ⅜ in. by ¹⁵⁄₁₆ in., leaving them slightly longer than the respective edge.

6. Band the curved edge first. Lay the substrate across a pair of bar or pipe clamps and use scraps that will raise it no more than ⅛ in. above the pipe. Set out the clamping caul and adjust the jaws of three additional clamps. Just as you did when gluing up the drawer fronts, apply glue to both sides of each veneer strip (except the last one) and bundle them up. Stand the bundle on edge against the substrate, move the caul into position, and apply the three clamps across the top of the panel. Tighten the center clamp first, then the clamps on either side of it, and finally the two outside clamps.

7. Trim the laminated band flush with the faces of the top and miter the ends after the

Tip: Veneer tape can eliminate the edge-gluing step. This special tape is so thin it won't dent the wood even under the force of the veneer press. Applied wet, it shrinks slightly as it dries, pulling the joint tight. Remove the tape with a cabinet scraper.

glue has cured. Because the front corners aren't 90 degrees, the miters aren't 45 degrees. Split the corner angle, marking a guide line across the band and onto the substrate surface. Saw through the band, just to the waste side of the line. Trim to the line with a chisel. You can clamp a straight-edged, fairly thick board right on the guide line to support the chisel as you pare.

8. Band the sides and back next. Miter the front end of each side band and make sure you have a tight joint. Mark the back end for mitering, cutting this to a 45-degree angle. Then glue and clamp the side edge bands to the top.

9. Next, fit and miter the back edge band. Glue it in place.

10. Trim the edge bands flush with the top.

11. Finally, cut and glue the gallery strip to the top surface a little shy of the back edge.

Photo H: Use a plunge router to cut the grooves for the inlay. Take the time to lay out the desired cuts and clamp stop blocks to the panel to control the beginning and end points. Guide the router along a straightedge.

Doing the side-panel inlay

The side panels are bordered with a thin inlay (as are the drawer fronts). You need to do this inlay work before assembling the sides. The inlay is a 1/16-in.-wide strip of contrasting wood. Edmundson used alder for the inlay on his chest. Maple or holly would work well, but other woods are viable alternatives.

1. Lay out the inlay grooves. The edge of the groove is 9/16 in. from the top and vertical edges of the side panel and 7/8 in. from the bottom edge.

2. Rout the grooves no more than 1/8 in. deep. Don't allow adjoining grooves to actually intersect (see **photo H**).

3. Connect the grooves at the corners with a tiny chisel. This is an exacting task that requires a sharp chisel and a steady hand.

4. Plane strips of contrasting wood for the inlay. Rip the strips as thin as possible and use a block plane to trim them to fit.

5. Decide how you'll join the corners of the inlay. The simplest method is to use a butt joint. That is, inlay the vertical strips, then fit the horizontal strips between them. More demanding is to miter the corners.

6. Glue the inlay into place. Cut a strip to the appropriate length and glue it into the groove. Cut the next strip and inlay it, and so on until all four strips are glued into place.

7. Plane or sand the inlay flush.

Assembling the case sides

Each side assembly consists of a pair of legs, a pair of side rails, and a side panel. The legs and rails are joined with the loose tenons, which you cut earlier. The panel is secured with biscuits. The biscuit slots still must be cut.

The faces of the rails are inset 1/8 in. from the faces of the legs. The face of the panel is inset another 1/8 in. from the faces of the rails. This means is that you have register the biscuit joiner from the inside faces of the rails and posts, since these faces are flush in the final assembly.

1. Lay out the biscuit slot locations on the rails and panel. Line up the parts, faces down, with the ends of the rails flush with the edges of the panel. Don't worry yet about the alignment of the faces. Just mark the biscuit locations on the panel and the rails.

2. Lay out the biscuit slot locations on the posts. Don't move the panel and rails. Simply move the posts into place, again with their faces down. Align the posts with their tops flush with the top rail. Make your marks on panel and posts.

3. Cut the slots in the panel, rails, and posts. These are straightforward cuts made with the biscuit joiner.

4. Glue up the side assemblies. Clamp them and set them aside until the glue cures.

Finishing the case sides

Next you need to glue the drawer guides to the side panels and do some routing. The drawer guides are glued to the inside of the side panels. Each guide fills the recess between the posts' inner faces and the panel, creating a continuous surface from front to back to guide the drawer's movement. The guides also get the grooves for the splines that join the intermediate web frames to the chest.

1. Thickness the drawer guide stock, leaving them a bit on the heavy side. Rip them to the width specified on the cut list and crosscut them to fit between the posts. Glue them to the side panels. Two or three screws can be used to clamp each guide. (If you dislike using fasteners, back the screws out after the glue sets.)

2. After the glue sets, withdraw the screws and handplane the faces of the guides flush with the posts.

3. Rout the spline grooves for the web frames. The grooves are ¼ in. wide and ⅜ in. deep and extend from the side assembly's back edge to within ½ in. of the front edge.

4. Cut the ⅜-in.-wide by ⅜-in. deep rabbet in the back posts for the case back. This can be cut with a router. The back itself overlays all the web frames, but it is rabbeted into the bottom back rail.

Making the bottom assembly

The bottom assembly consists of the bottom panel and the front and back rails.

1. Lay out the biscuit joints between the rails and the panel. The panel is to be flush with the top edges of the rails, so this work is straightforward. Lay the parts upside down on your workbench and butt them together as they'll be in the final assembly. Mark the biscuit locations on both the rails and bottom panel.

2. Cut the biscuit slots. This is a basic operation. Cut the slots in both the panel edges and the rails with the joiner square on the benchtop. The rails should be upside down, of course, when you cut them.

3. Shape the front edge of the bottom front rail. Use a template made from one of the drawer fronts.

4. Rabbet the back rail for the case back.

5. Glue up the rails and panel with biscuits. Clamp the assembly and set it aside to cure.

Making the web frames

The chest has four web frames (including the top one) and each is different. All are assembled in basically the same way, with rails and runners joined with loose tenon joints (see "Web Frame Construction" on p. 106). But two of the frames have center runners, whereas the others do not. Two of the frames are constructed of ⅝-in.-thick stock; the other two of ¾-in.-thick stock. The top frame has longer rails than the others to accommodate the tails that joint the frame to the case.

1. Cut the rails and drawer runners to the rough dimensions first, following the cut list specifications.

2. Dress the stock to the required thicknesses.

3. Crosscut rails and runners to their final lengths.

4. Cut a large single tail on each end of the front and back rails for the top web frame. The dimensions of these tails are shown in the drawing on p. 106. The cutting can be done with a jigsaw or on the bandsaw.

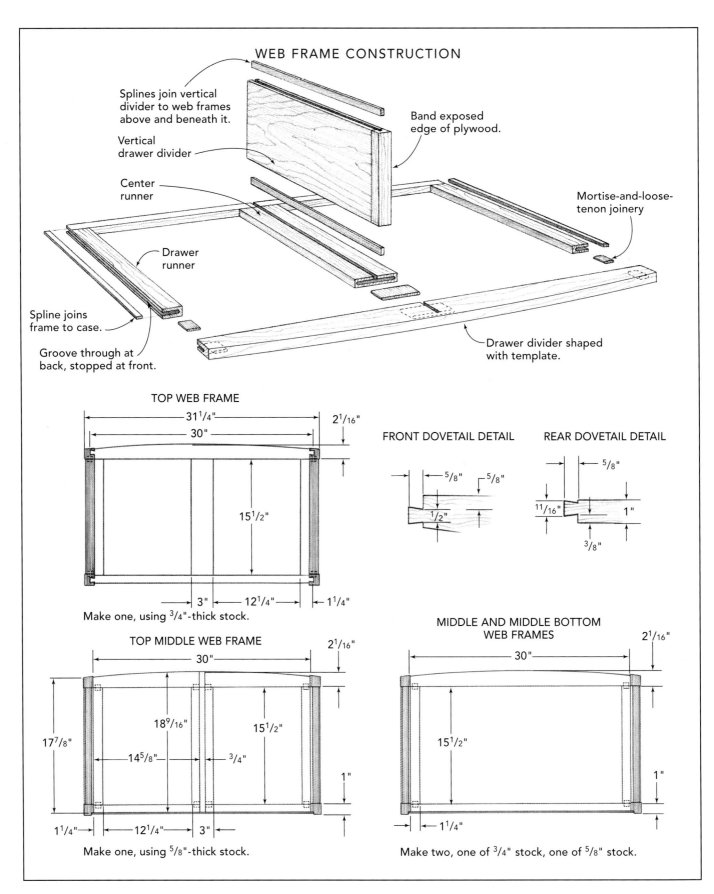

WEB FRAME CONSTRUCTION

Splines join vertical divider to web frames above and beneath it.

Vertical drawer divider

Center runner

Drawer runner

Spline joins frame to case.

Groove through at back, stopped at front.

Band exposed edge of plywood.

Mortise-and-loose-tenon joinery

Drawer divider shaped with template.

TOP WEB FRAME

31 1/4"
30"
2 1/16"
15 1/2"
3"
12 1/4"
1 1/4"

Make one, using 3/4"-thick stock.

FRONT DOVETAIL DETAIL

5/8"
5/8"
1/2"

REAR DOVETAIL DETAIL

5/8"
11/16"
1"
3/8"

TOP MIDDLE WEB FRAME

2 1/16"
30"
18 9/16"
15 1/2"
17 7/8"
14 5/8"
3/4"
1"
1 1/4"
12 1/4"
3"

Make one, using 5/8"-thick stock.

MIDDLE AND MIDDLE BOTTOM WEB FRAMES

2 1/16"
30"
15 1/2"
1"
1 1/4"

Make two, one of 3/4" stock, one of 5/8" stock.

5. Cut the mortises in both the rails and the runners. As a general guide, make the mortises ¼ in. wide, 1 in. deep, and ½ in. shorter than the runner's width.

6. Make a loose tenon for each joint.

7. Assemble each web frame. First run through an assembly without glue to check that everything fits and that the frame clamps up nice and square. If it does, glue up the frame.

8. Cut the front rail to shape using the template made from the drawer front. Use the same process you did to shape the bottom front rail.

Fitting the top frame

The top web frame is joined to the case with dovetails. To fit the frame properly, you need to assemble the sides and bottom (without glue). Do that, clamping the subassemblies so they are square; then get to work on the tails.

1. Set the top web frame on top of the posts. Make sure it is properly aligned.

2. Scribe around the tails with a marking or utility knife. Make sure you will be able to see the marks. Measure and mark the depth of slot necessary; then extend the knife marks down the inner surface of the post.

3. Chisel out the waste.

4. Carefully pare the slots until the frame can be seated. It should be flush with the tops of the posts.

Fitting the intermediate web frames

The intermediate web frames are joined to the case sides with splines. You've already grooved the drawer guides for these splines. Now you must slot the web frames and fit them into the case.

1. Slot the web frames. Do this on the router table, making the cut with the same bit used to groove the drawer guides on the case sides. The cuts are stopped at the front, but you haven't shaped the vertical drawer divider yet (see **photo I**). Measure the length of the

Photo I: Groove the web frames on the router table. Use a tall facing on the fence to better support the frames. You've got to tip the front of the frames back to begin the stopped cut in one edge. Then lift the frame up in the same way to end the cut in the other edge.

groove in the drawer guide and cut the same length groove in the web frames. Measure from the back edge and on the face of the drawer divider mark where the groove should end. You aren't going to be able to see the bit, of course, so stick a piece of masking tape on the router table next to the bit to mark exactly where it is.

2. Cut the splines and fit them to the slots.

3. Fit the web frames to the case. Just slide each frame into place. If you fit the splines in the web frame first, you can slide the frames into place from the back of the case. If you fit the slide to the case, the frames must go in from the front.

4. Measure between the top and top middle web frames for the vertical drawer divider, which is a piece of plywood with an edge band of primary wood. Slide it between the web frames, trimming the divider as necessary.

5. Rout a slot in the top and bottom edges of the vertical divider for the splines that will join it to the two web frames. Use the same router table setup you used to slot the web frames.

Shaping the drawer dividers

1. Lay out the ⅛-in. step from the post to the drawer divider. Mark each end of all the dividers.

2. Slide the web frames out of the case.

3. Use the template to shape the drawer dividers (see **photo J**).

4. Rout grooves in the top and top middle web frames for the vertical drawer divider. Use the same bit you used to groove the divider itself, but do this cut with a handheld router. Make sure you center the groove.

5. Reassemble the web frames to the chest and fit the vertical divider in place with splines.

Assembling the case

When you are sure that everything fits as it should, you can dismantle the chest in anticipation of gluing it together. When you assemble the chest with glue, continually check the chest to ensure that it is square. Don't install the back or the top until after the drawers have been made.

1. Join the side assemblies to the bottom assembly. Glue the loose tenons in the rails, then join the sides to them. Apply pipe clamps, but be sure to protect the wood.

2. Glue the top web frame in place.

3. Apply glue to the splines and slide the intermediate web frames into place.

4. Apply clamps across the chest, front and back, at each web frame. Again, protect the posts with cauls, so the metal clamp jaws don't dent them.

5. Glue the vertical drawer divider in place.

MAKING THE DRAWERS

The drawers in the chest are not your standard issue. Edmundson used a seldom-seen style developed in the early 1900s by Swedish manufacturer Nordiska Kompaniet. It's called, oddly enough, the NK drawer.

The NK drawer differs from the typical drawer in that it rides on shallow, wide slides, separate from the sides. The slides increase the bearing surface beneath the drawer, and reduce the contact between the sides and the case. The slides are cut first and joined to the drawer bottom; this assembly is carefully fitted to the case so it slides easily. Then an open-bottomed drawer box is constructed and mounted on the bottom assembly. An advantage here is that it's easier to fit the slides and bottom than the entire drawer.

The fact that the drawer fronts are bowed does add to the drawer-making challenge. You should already have the drawer fronts lami-

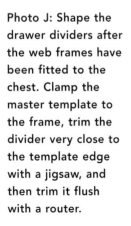

Photo J: Shape the drawer dividers after the web frames have been fitted to the chest. Clamp the master template to the frame, trim the divider very close to the template edge with a jigsaw, and then trim it flush with a router.

DRAWER CONSTRUCTION

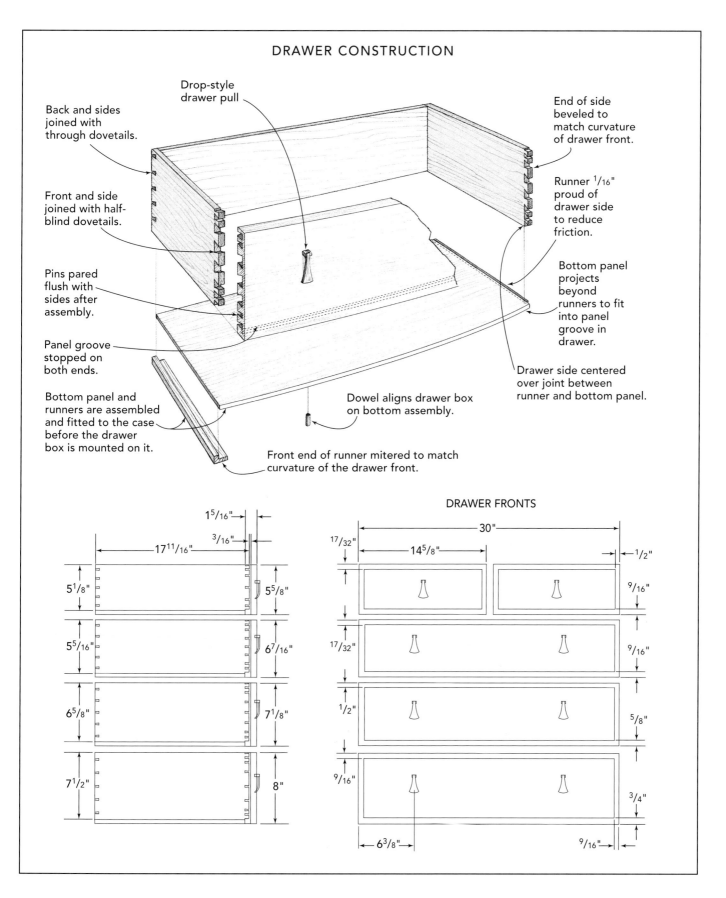

Drop-style drawer pull

Back and sides joined with through dovetails.

Front and side joined with half-blind dovetails.

Pins pared flush with sides after assembly.

Panel groove stopped on both ends.

Bottom panel and runners are assembled and fitted to the case before the drawer box is mounted on it.

End of side beveled to match curvature of drawer front.

Runner $^{1}/_{16}$" proud of drawer side to reduce friction.

Bottom panel projects beyond runners to fit into panel groove in drawer.

Drawer side centered over joint between runner and bottom panel.

Dowel aligns drawer box on bottom assembly.

Front end of runner mitered to match curvature of the drawer front.

DRAWER FRONTS

$1^{5}/_{16}$"

$^{3}/_{16}$"

$17^{11}/_{16}$"

$5^{1}/_{8}$" $5^{5}/_{8}$"

$5^{5}/_{16}$" $6^{7}/_{16}$"

$6^{5}/_{8}$" $7^{1}/_{8}$"

$7^{1}/_{2}$"   8"

30"

$^{17}/_{32}$" $14^{5}/_{8}$" $^{1}/_{2}$"

$^{9}/_{16}$"

$^{17}/_{32}$" $^{9}/_{16}$"

$^{1}/_{2}$" $^{5}/_{8}$"

$^{9}/_{16}$" $^{3}/_{4}$"

$6^{3}/_{8}$" $^{9}/_{16}$"

nated and awaiting your attention. But before you can work on them, you need to make the bottom assemblies. And even before you do that, you need to make a couple of templates to aid you in fitting some of the individual parts of the bow-front drawers.

Making fitting templates

To expedite the construction and fitting of the drawers, you need a couple of ¼-in. plywood templates. These are used in cutting the drawer fronts and in shaping the front edge of the drawer bottoms and slides. They need to extend from the front to the back of the drawer openings.

Photo K: The NK-style drawer has a separate bottom assembly consisting of a plywood panel and two slides. You need only a block plane to shave the sides of the slides to get the unit moving smoothly. Then you add the drawer box to it.

1. Use the convex master template to shape the front edges of the two fitting templates. Rout along the curved edge of the master with a pattern bit or flush-trimming bit to cut each of these fitting templates.

2. Fit one template to a full-width opening. Trim equal amounts from the sides to achieve a snug fit between the case side assemblies.

3. Cut the second template in half to fit it to the half-width drawer openings.

4. Trim the length of each template, aligning the back edge with the case back and account-

ing for the drawer front configuration. In this case, the drawer fronts are ⅝ in. thick and have a ¼-in.-deep groove for the bottom panel. For the drawer fronts to be flush with the drawer dividers, the templates should be ⅜ in. shy of the dividers.

Making the bottom assemblies

The NK drawer has a bottom unit consisting of the bottom panel and two slides. It is separate from the drawer box. The bottom assembly is cut, assembled, and fitted to the drawer opening before the box is even started.

1. Determine the dimensions of the slides. Edmundson made the slides for his chest ½ in. high and 1 in. wide. They need to extend from the inner face of the drawer front to the case back. Measure from the back edge of the web frame to the front edge of the drawer divider. From that measurement, subtract the thickness of the drawer front—⅝ in. This is the length of the slides. Because the front is curved, the front end of the slides must be shaped to mate with the fronts.

2. Cut the slides. Roughly size the slides; then use the fitting templates to mark the shape needed on the front ends. As with the front edges of the posts, a straight miter will do because the curve is so shallow.

3. Rabbet the slides for the bottom panels. The rabbet depth matches the panel thickness, and the width should place the panel's edge under the center of the drawer side. In this case, the side is ½ in. thick, and it is inset ¹⁄₁₆ in. from the outside edge of the slide. The rabbet thus should be ¹¹⁄₁₆ in. wide.

4. Set the slides in the drawer opening. Check the length; they should be ⅝ in. shy of the drawer divider's edge when they are flush with the case back. Measure between the rabbet shoulders to determine the width of the bottom panel.

5. Cut the bottom panels from plywood (or from the material you veneered earlier). Remember that the front edges must be shaped using the fitting templates and that the front edge should project ¼ in. beyond the front ends of the slides (see **photo K**).

6. Fit the parts in the drawer opening. Set the slides against the case sides and drop the bottom into the rabbets. Again, make sure the parts are square to the opening; the front ends of the slides should be the same distance from the drawer divider's edge. The assembly should be flush across the case back.

7. Glue the parts together. Make sure the back of the slides and the back of the panel are flush. A good way to clamp the parts uses the case itself. Apply the glue and assemble the parts inside the drawer opening. Set cauls on the bottom panel edges; then wedge sticks between the cauls and the runners above them.

8. Complete the fitting after the glue has set. Plane the sides of the slides as needed to get the assembly moving smoothly in and out of the case.

Fitting the drawer fronts

The drawer front blanks were glue laminated early on in the project. They should have been curing and stabilizing for several days at least. Now it's time to fit them to their respective drawer openings in the case. The pair of drawers on top will also be cut from a single glued-up blank at this time.

1. Clean up one edge of each drawer front with a scraper, then joint it. (You already did this to one drawer front.) Reference the fence with the convex face of the drawer front and feed the piece in an arc across the cutterhead. You want the edge straight. Make it that, and don't sweat getting it square. The table saw will square the edge.

2. Rip the fronts next. Keep the convex face down on the saw table as you cut. Clean up the unjointed edge with the first cut. This cut will make the edge straight and square to the face. Then turn the piece around and skim the jointed edge, bringing the piece to width and squaring the edge to the face. A slow but constant feed rate yields best results. The goal is a snug fit in the drawer opening. You can plane off some material to allow for wood movement later.

3. Lay out the slightly beveled crosscuts on the ends of the fronts. These bevels must be cut so that the edges end up parallel to the run of the drawer. To lay out the cutting line, use the fitting templates made earlier. Stand each front on edge, align the template on the exposed edge, and scribe along its sides onto the front. Use the full-width fitting template to lay out the end cuts on the top drawer front (which will soon be cut into two half-width fronts).

4. Make a cradle to hold the drawer fronts for crosscutting. It's nothing more than a pair of cleats fastened across the ends of a flat base. You set the drawer front face down on the cleats, and they support it and keep it from rocking. You set the cradle in a crosscut box used on the table saw. The base should be about 1 in. shorter and narrower than the largest front you will cut with it. The cleats should be 1 in. high and 1 in. to 1½ in. wide.

5. Crosscut the fronts. Make a first cut shy of the cutting line and make sure the cut is parallel to the line (see **photo L** on p. 112). If it is, shift the cradle and cut to the line. If the cut is not parallel to the line, you may need to raise one end of the front or the other with shims taped to the cleats.

6. Lay out the center cuts that separate the two top fronts from the lamination. Use the half-width template, aligning it flush with the end cuts, to mark the two cuts you need to remove a ¾-in.-wide strip in the middle of the front.

7. Add a pair of thin cleats to the crosscut cradle. These will support the last drawer front when you cut to the centerlines. Of course, you'll cut the cradle as well, but after these two cuts are completed, you are done with the cradle. Set the front in the cradle and plane down a strip of wood to just fit between the base and the drawer front's face on either side of the cut lines. Attach them to the base.

8. Cut the glue-up for the pair of top drawers in two. Do this the same way you cut the ends. Make a test cut shy of the line and make sure the cut is parallel to the line. Then shift the cradle, if necessary, and make the final cut.

Tip: Sometimes, in the heat of fitting drawers, you take off a little too much material and end up with a sloppy fit. With the NK system, you can just rip about ⅛ in. off the side of each slide, glue on a slightly thicker ripping, and start the fitting over.

Cutting the bottom groove

The bottom panel fits into a groove cut in the inside face of the drawer front. Because the drawer front is curved and because the groove has to be stopped on both ends, it's more challenging to cut than the through-and-through groove of the typical flat drawer front.

1. Make a simple setup gauge first. Set a bottom assembly on the benchtop and stand a piece of scrap next to its front edge. Mark along the top surface of the bottom panel on this scrap. This line delineates the top shoulder of the groove.

2. Set up the router table to rout the grooves. Use a slot cutter, but be sure to swap bearings, so the depth of cut is limited to ¼ in. (The standard cutting depth of a slot cutter is ½ in.) You must use the bearing, of course, because the curve of the front prevents the use of the fence. Use your gauge to set the height of the slot cutter. Stand the gauge next to the cutter and adjust it against the line.

3. Cut the groove. The cut is stopped on both ends, remember. Use a starting pin to begin

the cut, butting the work against the pin and pivoting it into the cutter (see **photo M**).

4. Use a chisel to square the ends of the groove, so the bottom will seat in it.

Making the drawer box

The box consists of the front, sides, and back. You cut the joints and assemble the parts to get the box ready to be mated to the bottom assembly.

1. Measure from the top of the groove to the top of the drawer front. This is the width of both the sides and the back.

2. Cut the sides and back. Rip them to width, but when you crosscut them, leave them several inches long. Until you join the sides and front, you don't know exactly how long the back must be. Leaving the sides long gives you the opportunity for a second shot at cutting the dovetails. If the joint doesn't fit to your satisfaction, you can cut off the tails and still have room to do them again.

3. Bevel the front end of each drawer side. You need to do this to mate the side properly to

Photo M: Making the slots in the drawer front stopped on both ends is the single biggest challenge in the operation. Use a starting pin to control the beginning of the cut and slide the work in an arc through the cut.

the curved front. Use the fitting template to mark each side, just as you did the ends of the drawer fronts. After marking them, cut the bevels.

4. Lay out and cut the dovetails (see "Cutting Dovetails on a Bow-Front Drawer" on p. 114).

5. Dry-fit the joints; then dry-fit the open-backed (for now) drawer box on its bottom assembly. Use this opportunity to establish the length of the sides and of the back. Determine the length of the back by measuring—just at the front—from the outside of one side to the outside of the other. Mark the sides along the back edges of the slides.

6. Dismantle the drawer. Trim the sides to the marked length. Cut the drawer back.

7. Lay out and cut the through dovetails to join the sides and back.

Assembling the drawer

Before gluing up the drawer, do one last dry-assembly. In the final assembly, you want the drawer front to line up perfectly with the outside edges of the drawer slides. If it overhangs one of the slides, it will, of course, prevent you from closing the drawer. And if you plane down that end of the drawer front, the appearance will be gappy on the other end. So make sure the front lines up properly. You also want the back centered on the bottom assembly, with the sides parallel to the slides. One good

CUTTING DOVETAILS ON A BOW-FRONT DRAWER

Cutting dovetails on curved drawer parts isn't terribly difficult. You just have to make a simple work holder for the curved fronts and a guide to hold your chisel on the same angle as the shoulder lines.

To hold the front when you chop the pins, make a concave bed like the laminating form (you could, in fact, use the form). Fine sandpaper glued to the bed creates an anti-slip surface.

To guide the chisel, crosscut a scrap at the same time that you bevel-crosscut the drawer sides. Set this board on the workpiece, its end parallel to the end of the workpiece and dead on the shoulder line, and clamp it.

To cut the pins on the drawer front, scribe around the front's ends with a marking gauge; then mark out the pins as usual with a sliding bevel gauge and pencil. Set the front in the bed and clamp it. Then saw the pins on a diagonal from scribe line to scribe line, just as you would standard half-blind dovetail pins. When chopping out the waste, clamp the chisel guide on the scribe line, its end parallel to the end of the front. Chop with the back of the chisel flat against the guide.

To cut the tails on the drawer sides, lay a drawer side down on the bench, then scribe the tails from the pins. Set a marking gauge to the pin depth and scribe the line for the depth of the tails. Angle the saw as you cut the tails, so the cut ends on the scribed lines. Butt the chisel is against the guide block when you chop out the waste. Chop halfway through the side, then flip it over. Finish chopping with the guide block oriented so its face is parallel with the side's end.

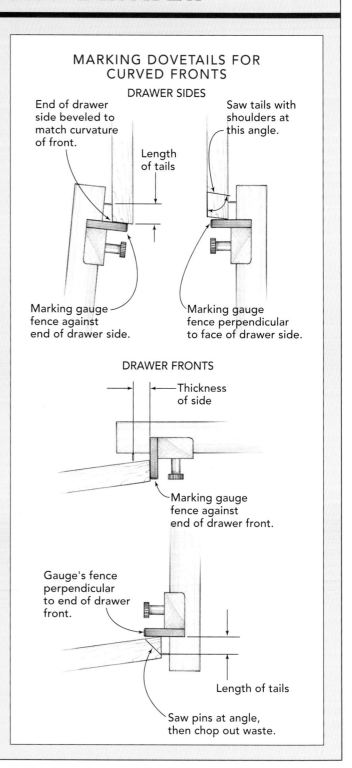

MARKING DOVETAILS FOR CURVED FRONTS

DRAWER SIDES

End of drawer side beveled to match curvature of front.

Saw tails with shoulders at this angle.

Length of tails

Marking gauge fence against end of drawer side.

Marking gauge fence perpendicular to face of drawer side.

DRAWER FRONTS

Thickness of side

Marking gauge fence against end of drawer front.

Gauge's fence perpendicular to end of drawer front.

Length of tails

Saw pins at angle, then chop out waste.

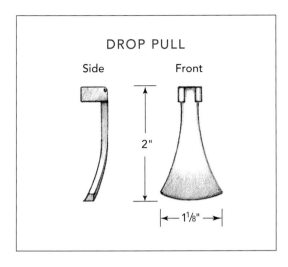

DROP PULL

Side   Front

2"

|← 1⅛" →|

way to ensure this happens is to worry this
detail when there's no glue setting and you can
take your time. Get it lined up; then drill a
hole though the bottom panel into the drawer
back. During the glue-up, you can drive a
dowel into the hole, forcing the assembly into
alignment.

1. Dry-fit the drawer, drilling the hole through
the bottom panel into the drawer back.
2. Glue up the drawer, at last. Apply glue to
the dovetails and assemble the drawer box.
Clamp the box; then continue with the assembly. Apply glue to the bottom edges of the
drawer box and to the groove in the drawer
front. Join the bottom assembly to the box.
Make sure the ends of the front align properly
with the slides. Drive the dowel in the alignment hole you drilled through the bottom.
Apply clamps and set the drawer aside until
the glue sets.
3. When the clamps have been removed,
check how the drawer fits in its opening.
Edmundson says that the need for minor
adjustments at this point is common. Typically,
though, only a bit of work with a handplane is
necessary to achieve a smooth-sliding fit.
4. Pare the pins flush. This can be done satisfactorily with a chisel or plane iron.

Inlaying the drawer front

The drawer fronts have inlay that matches the
side panel inlay. The materials, tools, and techniques used are the same. One difference is
that the drawer fronts are curved rather than
flat. Also different are the distances from the
drawer front edges to the inlay grooves.

Here is a list of the distances:
- **Top drawers:** ¹⁷⁄₃₂ in. from the top edge,
  ⁹⁄₁₆ in. from the bottom edge, ½ in.
  from the ends
- **Top middle drawer:** ¹⁷⁄₃₂ in. from the
  top, ⁹⁄₁₆ in. from the bottom, ⁹⁄₁₆ in. from
  the ends
- **Bottom middle drawer:** ½ in. from the
  top, ⅝ in. from the bottom, ⁹⁄₁₆ in. from
  the ends
- **Bottom drawer:** ⁹⁄₁₆ in. from the top, ¾ in.
  from the bottom, ⁹⁄₁₆ in. from the ends

Mounting the pulls

Edmundson designed a brass drawer pull
specifically for this chest and then made them
himself. There is, of course, a wide selection
of commercially made pulls (see "Sources"
on p. 216).

FINISHING UP

Only a few details remain. The back must be
glued in place, as must the top. Then you have
the finishing to do.

1. Measure the case for the back. It is flush
with the top surface of the top web frame.
2. Cut the back panel to final size.
3. Apply glue to the rabbets and to the back
edges of the web frames. Set the back in place
and clamp it.
4. Set the top in place and align it carefully,
flush with the faces of the back posts and with
an equal amount of overhang on either side.
5. Drill pilot holes up through the top web
frame at several locations and drive screws
through the holes.
6. Back out the screws, and remove the top.
Apply glue to the top frame, reposition the
case top, and redrive the mounting screws.
7. Apply the finish of your choice.

*Tip: Pare too deeply
into the dovetail
pins with a chisel,
and the end grain
can crumble. Try
using a plane iron
instead of a chisel.
Stack index cards
beside the pin—
all but flush with
top—and make
light shearing cuts.
Remove a couple of
cards at a time until
the pins are flush.*

DOUBLE DRESSER

Furniture makers often are reluctant to depart from tradition. They use solid wood throughout a chest, support drawers with web frames, and avoid using fasteners in favor of intricate joinery. The parts are crafted in ways that require a practiced hand.

Mark Edmundson, of Sandpoint, Idaho, isn't so tradition bound. He isn't reluctant to use modern materials, hardware, and joinery when circumstances recommend them. This dresser was commissioned by a couple with a budget too limited for expensive hardwoods and labor-intensive construction. To fulfill their desires within their means, Edmundson used veneered sheet goods, manufactured drawer runners, and biscuit and pocket-screw joinery.

None of this is immediately obvious. The style is spare and elegant. Though the figure of the wood is not showy, you don't immediately grasp that the dresser isn't solid wood throughout. From the front, it is all solid wood—leg posts, rails, drawer fronts, even the drawer pulls. No fastener or hardware is obvious, even when the drawers are open.

The result is a relatively simple project that can be duplicated with a minimum of shop equipment. You have to make the framework with considerable precision, but you use simple machine setups—rather than a practiced hand—to derive the accuracy needed.

Double Dresser

A CONTEMPORARY TAKE on traditional post-and-panel construction, the double dresser makes use of sheet goods panels, loose tenons, biscuits, and pocket screws.

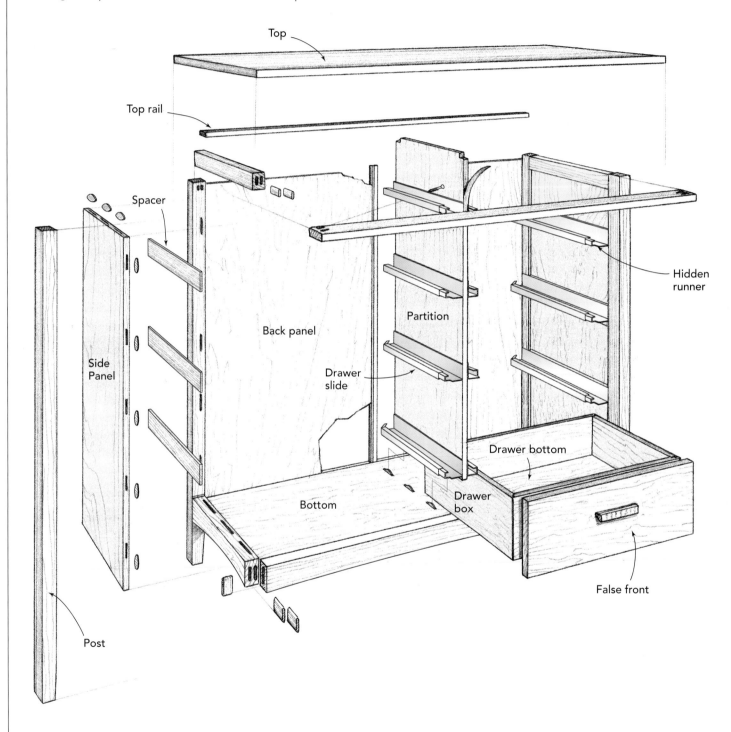

Top

Top rail

Spacer

Side
Panel

Back panel

Partition

Hidden
runner

Drawer
slide

Drawer bottom

Bottom

Drawer
box

False front

Post

SIDE VIEW

FRONT VIEW

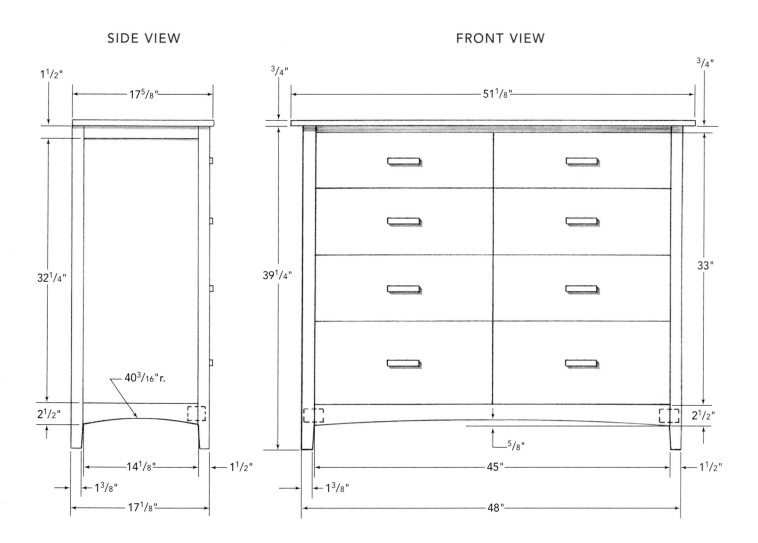

BUILDING THE CHEST STEP BY STEP

CUT LIST FOR DOUBLE DRESSER

Case

4	Leg posts	1½ in. x 1½ in. x 39¼ in.	maple
1	Bottom front rail	1⅜ in. x 2½ in. x 45 in.	maple
2	Bottom side rails	1⅜ in. x 2½ in. x 14⅛ in.	maple
2	Top side rails	1⅜ in. x 1½ in. x 14⅛ in.	maple
1	Top front rail	¾ in. x 1⅜ in. x 45 in.	maple
1	Bottom back rail	¾ in. x 1½ in. x 45 in.	maple
1	Top back rail	¾ in. x 1½ in. x 45 in.	maple
1	Partition edge band	¾ in. x ¾ in. x 33 in.	maple
2	Loose tenons	½ in. x 1¾ in. x 2½ in.	maple
2	Loose tenons	¼ in. x ¾ in. x 2½ in.	maple
8	Loose tenons	¼ in. x ¾ in. x ¾ in.	maple
8	Loose tenons	¼ in. x 1¾ in. x ¾ in.	maple
1	Bottom panel	¾ in. x 14⅛ in. x 45 in.	maple plywood
1	Partition panel	¾ in. x 16¼ in. x 33¾ in.	maple plywood
2	Back panels	¼ in. x 22⅝ in. x 33½ in.	maple plywood
2	Side panels	¾ in. x 14⅛ in. x 32¼ in.	maple-veneered MDF
2	Dowels	⁵⁄₁₆ in. diameter x 2 in.	
2	Long top edge bands	½ in. x ¾ in. x 51⅛ in.	maple
2	Short top edge bands	½ in. x ¾ in. x 17⅝ in.	maple
1	Top	¾ in. x 16⅝ in. x 50⅛ in.	maple-veneered MDF

CUT LIST FOR DOUBLE DRESSER

Drawers

2	Drawer fronts	¾ in. x 6¹¹⁄₁₆ in. x 22⅜ in.	maple
4	Drawer fronts	¾ in. x 8 in. x 22⅜ in.	maple
2	Drawer fronts	¾ in. x 10 in. x 22⅜ in.	maple
4	Drawer box front/backs	⅝ in. x 5 in. x 21¾ in.	alder
4	Drawer sides	⅝ in. x 5 in. x 14¾ in.	alder
8	Drawer box front/backs	⅝ in. x 6⅜ in. x 21¾ in.	alder
8	Drawer sides	⅝ in. x 6⅜ in. x 14¾ in.	alder
4	Drawer box front/backs	⅝ in. x 8⅜ in. x 21¾ in.	alder
4	Drawer sides	⅝ in. x 8⅜ in. x 14¾ in.	alder
8	Drawer bottoms	¼ in. x 14¼ in. x 21 in.	birch plywood
8	Runner spacers	½ in. x 3 in. x 14⅛ in.	plywood
8	Drawer pulls	¾ in. x ¾ in. x 4 in.	maple
16	Dowels	⁵⁄₁₆ in. diameter x 1 in.	

Hardware

8 pairs	Tandem single-extension drawer runners	15 in.	from Blum; item #552
8 pairs	Tandem locking devices		from Blum
16	Drawer front adjusters		from Blum

ONSTRUCTION OF THE DRESSER breaks down into two major phases: building the case and constructing the drawers. To build the case, you cut and join all the hardwood members into a framework. You then fill the openings with panels cut from sheet goods, joining them to the hardwood frame with biscuits.

CONSTRUCTING THE CASE

The case for the dresser is a contemporary take on the traditional post-and-panel construction. The legs and rails are made of hardwood, and they are assembled with loose tenons, a modern form of the mortise-and-tenon joint. But the panels are stable, man-made materials, and they are installed with easy-to-cut biscuit joinery.

Preparing the hardwood stock

The legs, the lower front rail, and all the side rails are made from 8/4 hard maple. The top front and both back rails are made from 4/4 hard maple, as is the rear edge band on the partition.

1. Rip and crosscut the maple parts to rough sizes, as appropriate, before dressing the stock.
2. Joint a face and an edge of each piece.
3. Plane the stock to the correct thickness. The legs are all 1½ in. square. If you cut individual blanks from the rough stock, you can make them the final width and thickness using the thickness planer. The bottom front rail and the side rails are 1⅜ in. thick. All the parts taken from the 4/4 stock dress out to ¾ in. thick.

Mortising the legs and rails

Edmundson used mortises and tenons to join most of the rails to the legs. His joint configurations are not entirely what you might expect:

- The front bottom rail has a large single tenon, which penetrates the leg as far as possible (1¼ in.).
- The side rails are joined to the legs with short twin mortise-and-tenon joints. The mortises for them are as deep as they can be (⅜ in.) without penetrating the mortise for the front rail.
- The bottom back rail is narrower and thinner than the front rail, and its mortise is too. But it penetrates the leg the same distance as the front rail's mortise (1¼ in.).

All these joints use a loose, or slip, tenon. To make this joint, you cut mortises in both mating pieces; then join the pieces with a separate tenon that is glued into both mortises. Since you will probably use the same tool to do all the mortises, cut them together. (Edmundson uses a horizontal boring machine.)

"Case Joinery" shows the dimensions and placement of these joints.

1. Cut the mortise in the leg for the bottom front rail. It is ½ in. wide, 1¾ in. long, and 1¼ in. deep. Lacking a horizontal borer, use a router or hollow-chisel mortiser.
2. Cut the matching mortises into both ends of the front rail. The length of the rail makes the staging a bit of a challenge if you do this task with a router, and especially with a hollow-chisel mortiser.
3. Next, change your tool setup to cut ¼-in.-wide mortises.
4. Begin by cutting the deep mortise in the back legs for the bottom rail.
5. Cut matching mortises into the rail ends.
6. Now cut the twin mortises. Positioning the mortises consistently is critical. With the router setup I use (see "Mortising Jig" on p. 124), you can do both of the twins with the same setup. You simply move a spacer to shift the router on the jig.

Tapering the feet

Each leg is tapered slightly. The taper—amounting to just ⅛ in.—extends from the bottom of the lower rail to the foot, a distance of 3 in. It is cut on only the two inside faces (beneath the rails). The adjacent outside faces are straight from top to bottom.

1. Lay out the tapers on the legs. Mark the shoulder of the lower rails on the legs. With a rule and pencil, mark the cut line from the shoulders to the foot.
2. Handplane the tapers. This proves to be much quicker and easier and less risky to do than sawing them.

Shaping the bottom rails

The bottom edge of the lower rails on the sides and the front are arched, rising ⅝ in. from the ends to the center point.

1. Lay out the cut lines on the side rails. This is relatively easy, because the side rails are short. Mark the center point and measure ⅝ in. up from the bottom edge at that point. Have a helper flex a thin batten so it arches from one bottom corner to the center mark and back to the opposite bottom corner. Scribe a line along the flexed batten.
2. Lay out the cut line on the front rail, using the same procedure.
3. Cut to the lines on the bandsaw.
4. Sand the cut edge to smooth it. This is easiest to do with a spindle sander or at the drum of an edge sander.

Grooving for the back panels

The back panels are housed in grooves plowed in the legs, the top and bottom back rails, and the partition edge band. For the best fit, you should cut this groove on the router table so you can reference the cut off the back face (or edge) of each part.

1. Measure the thickness of the plywood you're using for the back panels. If the material is dramatically shy of ¼ in., you may want to use a ⁷⁄₃₂ in. bit to plow the grooves. Just

CASE JOINERY

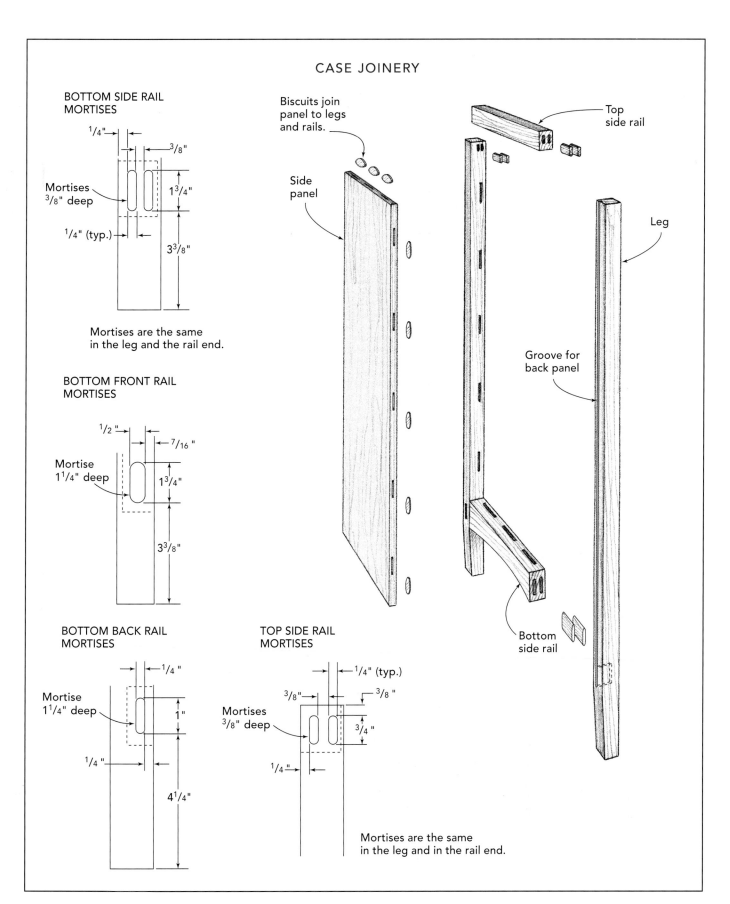

BOTTOM SIDE RAIL MORTISES

1/4"
3/8"
Mortises 3/8" deep
1 3/4"
1/4" (typ.)
3 3/8"

Mortises are the same in the leg and the rail end.

BOTTOM FRONT RAIL MORTISES

1/2"
7/16"
Mortise 1 1/4" deep
1 3/4"
3 3/8"

BOTTOM BACK RAIL MORTISES

1/4"
Mortise 1 1/4" deep
1"
1/4"
4 1/4"

TOP SIDE RAIL MORTISES

1/4" (typ.)
3/8"
3/8"
Mortises 3/8" deep
3/4"
1/4"

Mortises are the same in the leg and in the rail end.

Biscuits join panel to legs and rails.

Top side rail

Side panel

Leg

Groove for back panel

Bottom side rail

MORTISING JIG

This jig is designed for small-scale production mortising. The work holders adjust for different workpiece sizes (and orientations). The bit diameter establishes the mortise width, while the router's plunge controls the depth. The track and adjustable stops control and limit the router's movement. Mark only the midpoint for a mortise; align that mark with the registration line, and the jig and router will produce the mortise.

When cutting the mortise, the router is on top of the jig, with its edge guide riding along the jig's back edge. The stops control the length of the mortise by arresting the router's movement.

Use the vertical work holder when cutting end mortises. Hold the workpiece against the face of the jig with the toggle clamp, but also apply a clamp to cinch the piece against the work holder.

Switch to the horizontal work holder to rout side mortises. The setup of the router and the jig's stops shouldn't have to be changed. Moving the spacer in front of or behind the edge-guide facing in the track bumps the router position back and forth for doing twin mortises; no other changes are needed.

MORTISING JIG PLAN

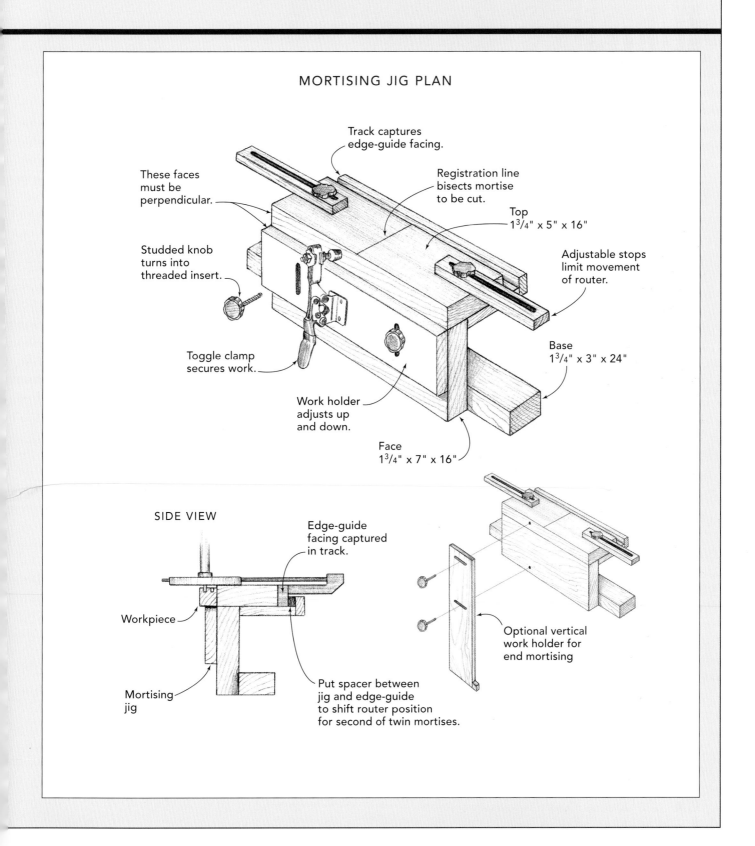

Track captures edge-guide facing.

These faces must be perpendicular.

Registration line bisects mortise to be cut.

Top
1³/₄" x 5" x 16"

Studded knob turns into threaded insert.

Adjustable stops limit movement of router.

Toggle clamp secures work.

Base
1³/₄" x 3" x 24"

Work holder adjusts up and down.

Face
1³/₄" x 7" x 16"

SIDE VIEW

Edge-guide facing captured in track.

Workpiece

Mortising jig

Put spacer between jig and edge-guide to shift router position for second of twin mortises.

Optional vertical work holder for end mortising

Photo A: Locate back panel grooves consistently in the legs, partition edge band, and rails by referencing the router table fence with the "back" surface of the workpiece.

bear in mind that you do have to slide the panel into place. Don't let a compulsion to avoid an overly loose fit, steer you into assembly problems.

2. Set up the router table. First, secure the bit in the collet. Set the cut depth to about ⅛ in. (The stock is hard and a ¼-in. bit is relatively frail, so cut to the full depth of ¼ in. in two passes.) Position the fence so the groove will be ¼ in. from the reference face of the work.

3. Groove the parts. In all but the legs, the grooves extend end to end. In the legs, they extend from the top down to the mortise (see **photo A**). The partition edge band gets two grooves. Cut each groove in two passes.

Making the loose tenons

1. Mill scraps of the maple to the thicknesses needed. If you use power tools for this, use scraps that are long enough to be machined safely. Be sure the material is wide enough for the applications, of course. The front rail mortises require ½-in.-thick stock; all the others require ¼-in.-thick stock.

2. Rip the stock to the needed widths.

3. Round the edges of the tenon stock with on the router table. Use either bullnose or roundover bits.

4. Crosscut the needed tenons to dimensions shown in the cut list.

Dry-assembling the legs and rails

Even if you can't assemble all the frame parts at once, you can join selected ones into subassemblies. The point of the exercise is to make sure everything fits, of course, and that the resulting assemblies are square and flat. But you also want to measure for the panels. (You don't want to glue up the bottom panel and rails only to discover that the mortises don't line up.)

1. Slip the tenons into mortises in the side rails (see **photo B**).

2. Fit the legs onto the tenons.

3. Clamp the assembly. If you have metal-jawed clamps that might mar the work, use cauls between jaws and wood.

Photo B: A loose tenon is easy to fit snugly in its mortise. Prepare a long strip to the proper thickness and width, bullnose the edges on the router table, and crosscut individual slips.

4. Check the assembly for squareness and flatness.

5. Measure the panel space carefully and note the dimensions.

6. Assemble the second side in the same way. Note the dimensions of its panel space.

7. Insert the tenons into the front and back rails.

8. Join the rails to the two clamped-up side assemblies (see **photo C**). Apply the clamps so they parallel the rails.

9. Check the assembly for squareness and flatness.

10. Measure the space for the bottom panel and note the dimensions.

Cutting the panels

Edmundson used both veneered MDF and hardwood plywood in the dresser. The MDF was used for the end panels and the top, where the plain-sliced veneer would look best. The plywood was used in the bottom and the back, and for the divider, places that typically aren't seen.

Photo C: A dry run allows you to check the fit of the joints and determine the dimensions for the side, bottom, and back panels. Clamp the sides; then join them to the front and back rails.

Tip: *To avoid chipping the veneer when cutting the MDF, rip the panels to a rough width first. After crosscutting to the final length, rerip the panel to final width. In doing so, you'll cut away the ragged corners left from the crosscuts.*

1. Cut the bottom panel to the dimensions you took from the dry-assembled case framework.

2. Cut the partition panel from the same sheet of plywood.

3. Cut the two back panels.

4. Cut the two side panels from ¾-in. A1 plain-sliced maple-veneered MDF.

Making the bottom assembly

The bottom unit consists of the bottom panel and the bottom front and back rails.

1. Lay out the biscuit joints between the rails and the panel. The panel is to be flush with the top edges of the rails, so this work is straightforward. Lay the parts upside down on the workbench and butt them together as they'll be in the final assembly. Mark biscuit locations on both rails and panel.

2. Cut the biscuit slots. This is a basic operation. Cut the slots in both the panel edges and the rails with the joiner square on the bench-

top. The rails should be upside down, of course, when you cut them.

3. Glue up the rails and panel with biscuits. Clamp the assembly and set it aside to cure.

Making the side assemblies

Each side assembly consists of a pair of legs, a pair of side rails, and a side panel. The legs and rails are joined with the mortise-and-loose-tenon joints, which have already been cut. The panel is secured with biscuits, and their slots must be cut now.

The only departure from the straightforward is the progression of insets. The faces of the rails are inset ⅛ in. from the faces of the legs. The face of the panel is inset another ⅛ in. from the faces of the rails. All this means is that you have to settle on a consistent way to cut the slots in the rails and legs so they'll align with the slots in the panel edges.

1. Lay out the slot locations on the rails and panel. Line up the parts, faces down, with the ends of the rails flush with the edges of the

Photo D: Use the biscuit joiner's fence to locate the side panel biscuit slots properly in the legs and rails. By referencing from the inside of the assembly, you can use the same setting when slotting both rails and legs.

Photo E: Glue tenons and biscuits into the legs, and biscuits into the panel ends. Fit a rail and the panel to a leg. Add the second rail, and the second leg.

panel. Don't worry yet about the alignment of the faces. Just mark the biscuit locations on the panel and the rails.

2. Lay out the slot locations on the legs. Don't move the panel and rails. Simply move the legs into place, again with their faces down. Align the legs with their tops flush with the top rail. Make your marks on the panel and on the legs.

3. Cut the slots in the panel. This is a straightforward cut with the biscuit joiner.

4. Cut the slots in the rails and legs. The backs of the legs and rails are flush, so use them as the reference surface when cutting the slots. Center the slots ⅞ in. from the back edge of these parts. Use the joiner's fence to locate the slots and to orient them square to the faces (see **photo D**). And make sure the rails are face down when you make the cuts.

5. Glue up the side assemblies (see **photo E**). Clamp them and set them aside until the glue cures.

Making the partition

The partition panel divides the case into two bays and supports the drawer runners. The partition, which was cut earlier, is banded on its back edge with a hardwood strip that's already been grooved for the back panels. Its front edge is banded with iron-on veneer tape. That edge is hidden by the drawer fronts in the completed dresser (see "Partition Joinery" on p. 130).

1. Glue the partition edge band to the partition panel's back edge.

2. Cut the notches for the top rails. The edge of the back rail is flush with the back edge of the partition, so the notch must be 1½ in. wide. Cut it ¾ in. deep. Because the partition

PARTITION JOINERY

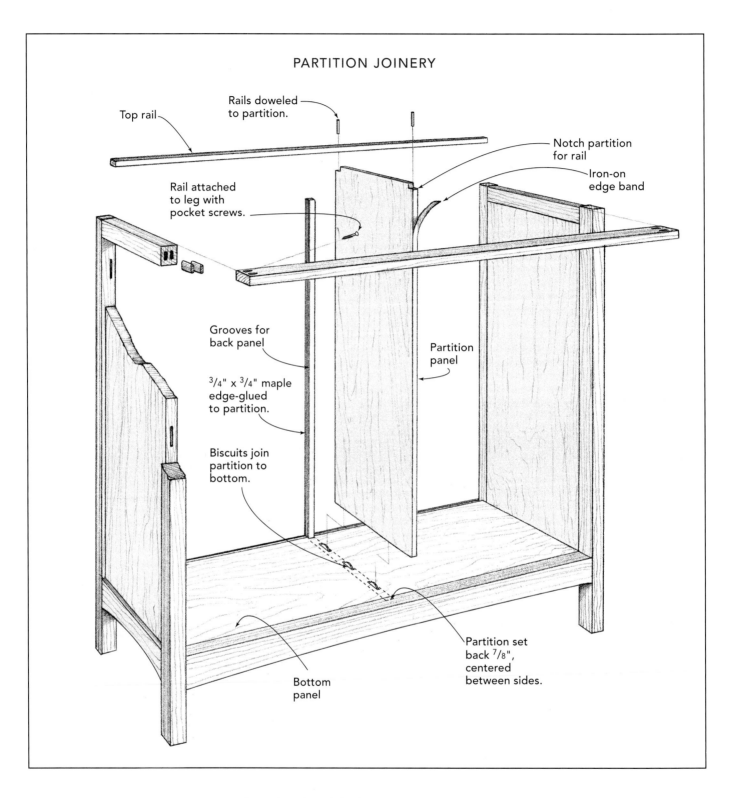

Top rail

Rails doweled to partition.

Notch partition for rail

Iron-on edge band

Rail attached to leg with pocket screws.

Grooves for back panel

Partition panel

$3/4" \times 3/4"$ maple edge-glued to partition.

Biscuits join partition to bottom.

Partition set back $7/8"$, centered between sides.

Bottom panel

is set in ⅞ in. from the edge of the front rail, the notch for the front rail need only be ⅝ in. wide. It too is ¾ in. deep.

3. Apply the iron-on veneer tape to the front edge of the partition panel. If you don't have

veneer tape, just glue a thin (⅛-in. or less) maple ripping to the edge.

4. Cut biscuit slots in the partition panel and the bottom assembly. Stand the partition on the bottom assembly, center it, and align the

Photo F: Glue biscuits in the slots in the case bottom and maneuver the partition into place, nestled under the front rail.

back edge flush with bottom's back edge. Mark the slot locations. Lay it flat on the bottom and cut the slots in it and in the bottom.

Assembling the case

1. Join the side assemblies and the bottom assembly, using glue. Apply clamps.

2. Attach the top front rail to the legs with pan-head pocket screws. Clamp scraps to the legs to support the rail, flush with the top of the legs, while you drive the screws.

3. Glue the partition panel to the bottom. Glue the biscuits into the slots in the bottom and slip the partition into place (see **photo F**). You need to dip the notched front corner under the top rail as you engage the biscuits. Align the panel flush with the back edge of the case.

4. Drill a 2-in.-deep hole for a ⁵⁄₁₆-in.-diameter dowel through the front rail into the edge of the partition.

5. Glue the dowel in the hole.

Installing the back panels

Although the back of the chest is plywood, it has a more finished look than the usual plywood back. Instead of nailing the plywood into a rabbet, the two panels are housed in grooves in the legs and rails. The panels and grooves have already been cut.

1. Glue the back panels in place. Edmundson used polyurethane glue because it lubricates better than yellow glue. Run a bead of it in the grooves for one panel; then slide that panel into place. Repeat to install the second panel.

2. Install the top back rail. Run a bead of polyurethane glue in the groove in the top back rail. Set the rail in place and drive the pocket screws that secure it to the legs (see **photo G** on p. 132). Working quickly and thoroughly, use mineral spirits to clean any squeezed-out glue from the case.

3. Dowel the rail to the partition. Drill a ⁵⁄₁₆-in.-diameter by 2-in.-deep hole through the

Tip: MDF edges
absorb glue readily.
So it is easy to get
a glue-starved joint.
To prevent this, size
the MDF edge with
glue. Spread glue
on the edge and
allow it to stand for
about 10 minutes.
Then reapply glue
and complete the
glue-up.

Photo G: Mount the top back rail as soon as the back panels are seated. Get a corner of one panel started in the rail's panel groove, and lower the rail into place.

rail into the edge of the partition. Glue a dowel in the hole; then trim it flush.

Making and attaching the top

The top is a piece of maple-veneered MDF to which you apply maple edge bands.

1. Prepare the maple edge bands for the top. Mill 4/4 stock to $^{13}/_{16}$ in. thick and rip $^{1}/_{2}$-in.-thick strips from it.

2. Cut the top panel.

3. Fit the edge bands to the top, mitering the ends of each piece.

4. Glue the bands in place.

5. Plane the edge bands flush with the top and bottom of the panel.

6. Position the top, with its back edge flush with the back surfaces of the legs and back rail. Drive screws up through the top rails into the top.

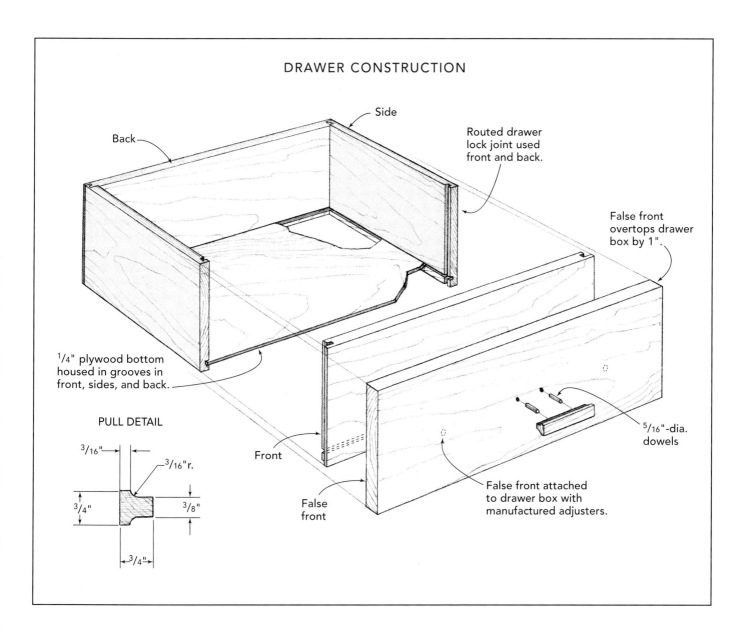

DRAWER CONSTRUCTION

Back

Side

Routed drawer lock joint used front and back.

False front overtops drawer box by 1".

1/4" plywood bottom housed in grooves in front, sides, and back.

PULL DETAIL

3/16"

3/16" r.

3/4"

3/8"

3/4"

Front

False front

False front attached to drawer box with manufactured adjusters.

5/16"-dia. dowels

CONSTRUCTING THE DRAWERS

All the drawers are solid-wood boxes with enclosed plywood bottoms and maple false fronts attached with adjusters. The pulls are made from tiger maple and mounted to the false fronts with dowels (see "Drawer Construction").

Edmundson mounted the drawers on hidden runners. Key dimensions are affected by this hardware. The length of the drawer boxes is dictated by the 15-in. length of the runners. Beneath each drawer, ⅝ in. of clearance is

required for the tracks. The box must be ⅜ in. narrower than the drawer opening. To provide space for the locks that clip the drawer to the runners, a ½-in. space is needed between the drawer bottom and the bottom edge of the drawer sides.

Making the drawer boxes

Any wood can be used for the drawer boxes, including plywood. Edmundson used alder because it is common in his area.

1. Cut the parts. Aim for tolerances of ½₂ in. when working with the hidden runners.

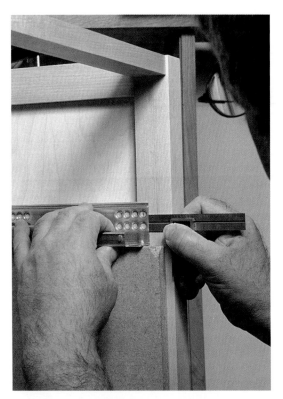

Photo H: To locate the runner hardware on the case sides and partition, use a plywood jig to hold it at the correct elevation while you drive the mounting screws.

2. Cut the bottom groove. For the best fit, you want the grooves just wide enough for whatever plywood you use for the bottom. Doing the operation on the table saw allows you to fine-tune the groove width. Make the initial cut ½ in. from the bottom edge of each front, back, and side. Adjust the rip fence and make the final pass.

3. Cut the joinery. Edmundson used a drawer-lock router bit in the table-mounted router to cut the joints. Setting up requires a few test cuts to fine-tune the bit height and the two fence settings—one for cutting the sides and one for cutting the fronts (see "Routed Drawer Lock Joint"). Once that the settings are dialed in, cut the joints.

4. Assemble the drawer boxes. The bottom should force the drawer into square, but work on a surface you know is flat to avoid twist.

Mounting the drawer hardware in the case

The drawer runners are integral to the mounting brackets, which you are about to screw to the case sides and partition. To fit a drawer in place (once the hardware is mounted), you set it atop the runners and slide it toward the back. At the drawer back, a hooked tab on the runner is engaged; and at the same time, a pair of clips under the drawer engage the ends of the runners. The clips connect the runners and drawer so they move together. To remove the drawer, you open the drawer partway, disengage the clips, and pull the drawer toward you. Comes right out.

To locate the runner hardware so the four runners for each pair of drawers are mounted at the same height, use a simple plywood jig. Begin with the case-mounted hardware.

1. Cut eight spacers for the hardware that mounts to the case sides. You need to have the hardware flush with the surfaces of the legs that face the partition. Thickness your spacers so they will do that.

2. Cut a scrap piece of plywood to use in locating the drawer runner hardware in the case. Cut it to the depth of the case and the height of the top track location. Mark one corner. When you use this jig, always position the marked corner to the bottom and back of the case. That will ensure that all the tracks will be oriented consistently.

3. Mount the first track unit. Set the jig into the case, rest it on the bottom, and hold it against the case side. Fit a spacer between the legs and fasten it. Set a track unit on the plywood and adjust its front-to-back alignment. Screw it to the case (see **photo H**).

4. Move the plywood jig across the drawer bay and hold it against the partition. Set a track unit on top of it, against the partition. Adjust the front-to-back alignment; then screw it to the partition.

5. Install the top two tracks in the second drawer bay in the same manner.

6. Cut down the jig and use it to mount the next tier of runners. Cut it again and mount the third tier. The lowest runners rest on the case bottom and no jig is needed.

ROUTED DRAWER LOCK JOINT

Fast, secure drawer construction is the purpose of this joint. It doesn't have the pizzazz of dovetails, but it can be cut a whole lot faster.

The joint is cut with a single router bit used in the table-mounted router. Once the height of the bit is properly set—do test cuts to dial in the setting—you use it for both halves of the joint. You lay the drawer fronts (and backs) flat on the tabletop to cut them. You stand the sides on edge against the fence to cut them, as shown. Only the fence position changes between these cuts.

The joint can be used with any thickness of stock and any mixture of stock thicknesses. It will produce flush drawers and lipped drawers.

Stand a drawer side on end, its inside face flat against the fence, to rout it. Featherboards eliminate workpiece bobbles.

SETTING THE FENCE

FOR DRAWER FRONT

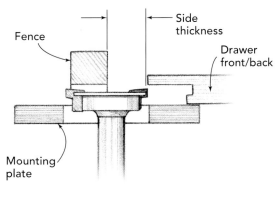

Fence

Side thickness

Drawer front/back

Mounting plate

FOR DRAWER SIDES

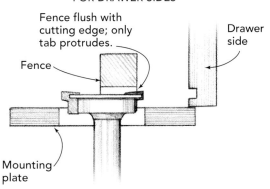

Fence flush with cutting edge; only tab protrudes.

Fence

Drawer side

Mounting plate

Lay the drawer front (or back) flat on the tabletop. Guide it along the fence with a push block when making the cut.

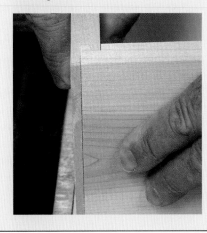

The assembled joint has a strong mechanical lock and expanded glue area.

Photo I: Latches fas-
tened underneath
the drawer clip it to
the runners. A jig
from the manufac-
turer guarantees
proper alignment
of the pilot holes
for the mounting.

Mounting the hardware on the drawers

As noted, the runners can't be removed from the case-mounted brackets. Instead, you extend them, set the drawer onto them, and latch them. Now you are going to notch each drawer back to accommodate the runners, and mount a pair of latches on each drawer that will hook it to the runners.

1. Screw the latches (called "locking devices" by Blum) to the underside of each drawer. The latch is set on the bottom, tight against the drawer front. Drill pilot holes at the mounting points and screw the latches in place. The mounting holes in the latches can guide the drill bit, but jigs and special drill bits are available, and they do help (see **photo I**).
2. Notch the drawer backs below the bottom for the runners. Follow the manufacturer's

instructions regarding size and exact place-ment of the notches.
3. If necessary, drill holes in the drawer backs. Some brands of hidden slides have a hook on the runner that captures the back to help sta-bilize the drawer. If the runners you select have this feature, drill the hole. As with the latches, a jig is available from the manufac-turer to make this task easy.

Mounting the drawer fronts

The drawer fronts are mounted to the boxes by way of commercial adjusters, which allow you to fine-tune the alignment of the front after it is attached to the drawer box.

1. Drill shallow, flat-bottomed holes in the backs of the drawer fronts for the adjusters.
2. Work out the alignment of each drawer front on its drawer in terms of the amount of overhang on each side.
3. Attach the drawer boxes to the fronts. Set the front on the workbench, position the box on it with the correct overhang, and drive mounting screws through the box into the adjusters, following the manufacturer's instructions.
4. Put the drawers in the case. Use the adjusters to refine the alignment of the fronts, so you have a uniform grid of fronts.

Making the drawer pulls

Each drawer has a single 4-in.-long pull mounted in the middle. The pulls can be made from scraps of the maple used to make the drawer fronts. Edmundson used tiger maple.

1. Cut several 12-in.- to 18-in.-long strips of ¾-in. by ¾-in. maple.
2. Using a ⅜-in.-diameter corebox bit in the table-mounted router, plow a ³⁄₁₆-in.-deep groove in the center of opposing faces of each strip.
3. Switch to a straight bit and plow a ⅜-in. wide by ³⁄₁₆-in.-deep rabbet into the same faces, trimming away one shoulder of the groove and transforming the section of each strip into a T-shape.

Photo J: Drill dowel holes in the back of each pull. A simple fixture clamped to the drill-press table ensures the holes are located consistently.

4. Sand the pull strips as necessary to blend the surfaces and soften hard edges.

5. Crosscut the strips into 4-in.-long pieces, making one for each drawer.

6. Make a fixture to use in drilling two holes in each pull for the 5⁄16-in.-diameter mounting dowels. Set a pull on a scrap of 3⁄4-in. plywood and nail trap fences all around it. Align the fixture and pull on the drill-press table so the first hole will be exactly where you want it. Clamp the fixture to the table.

7. After setting the drill press for the depth of hole you want, drill a hole in the pull. Lift the pull out of the fixture, turn it around, and return it to the fixture. Drill the second hole (see **photo J**). Drill holes in all the pulls in this way.

8. Before breaking down the fixture used to hold the pulls for drilling, use it to bore holes completely through a scrap of hardwood. Use this as the basis for a jig to guide you when you drill mounting holes for the pulls in the drawer fronts.

9. Lay out the locations for the pulls on the drawer fronts. Lightly scribe a horizontal line across each drawer front—the line on which

you'll center the pull. Bisect the line with a perpendicular.

10. Make a jig with the drilling guide block. Align it on the layout lines and drill mounting holes for a pull on each drawer front.

11. Apply glue sparingly to short pieces of dowel. Drive a dowel into each mounting hole in the drawer front, then press a pull onto the projecting ends of the dowels.

FINISHING UP

The final task is to apply your favorite finish to the dresser. Edmundson used a clear finish on his dresser, and I used a brush-on polyurethane on mine. For the most thorough job, remove the drawer fronts from the drawer boxes and apply the finish to the back as well as to the front. I left the drawer boxes unfinished. Remove the runners from the case and finish the case inside and out equally. Then remount the fronts and the runners, slip the drawers in place, and fill the dresser with clothing.

Tip: Routing the rabbet that forms the drawer pulls removes most of the support for the work. To support the work, stick a 3⁄16-in.-thick by 1⁄4-in.-wide strip to the tabletop with a couple patches of double-sided tape. Position it on the outfeed side of the bit, against the fence.

TRIPLE DRESSER

It is big—6 ft. long and nearly 3½ ft. high. It's complicated too—a mammoth solid-wood case with hand-cut dovetails, four web frames, two face frames, a frame-and-panel back, a leg-and-apron base, and 12 drawers in eight different sizes.

The triple dresser blends the design style of the furniture maker Michael Seward and the configuration his customer had in mind. His client wanted the three-drawers-across arrangement, and she also wanted drawers for her jewelry. A small chest on the dresser top would interfere with a wall-hung mirror, so Seward suggested incorporating shallow drawers in the dresser itself.

The big engineering challenge was to provide support for three banks of drawers. The case has no vertical dividers. Each web frame is nearly 6 ft. long and supports three drawers. To keep the frames from sagging, Seward incorporated a second face frame, this one at the back of the case. Though slender, two stiles provide just the support needed, and they don't reduce the depth of the case.

Seward, from York County, Pennsylvania, was influenced initially by the designs of Thomas Moser. His designs tend to be low key, but the material he uses is not. Inevitably, what hits you first is the figure of the wood. When you look closer, you see the meticulous craftsmanship.

Triple Dresser

THE TRIPLE DRESSER is a large but simple box made of four wide boards, joined with dovetails. Most of its 12 drawers are supported on web frames (3 drawers ride on the case bottom). A face frame in front and a support frame in back brace the web frames. A frame-and-panel assembly closes in the back. The entire construction rests on a stout leg-and-rail base.

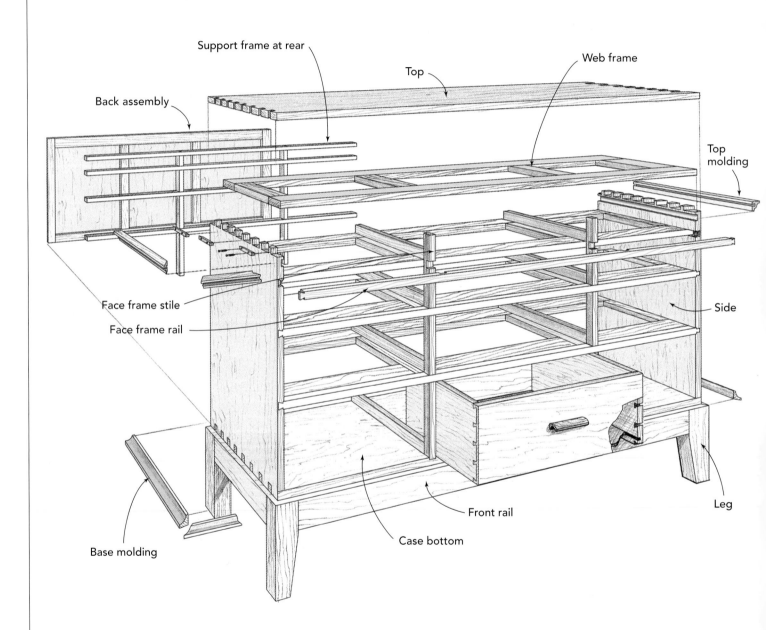

Support frame at rear

Top

Web frame

Back assembly

Top molding

Face frame stile

Face frame rail

Side

Base molding

Front rail

Case bottom

Leg

BACK VIEW

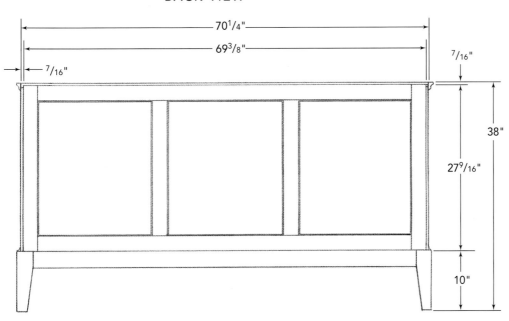

70$\frac{1}{4}$"

69$\frac{3}{8}$"

$\frac{7}{16}$"

$\frac{7}{16}$"

38"

27$\frac{9}{16}$"

10"

FRONT VIEW

END VIEW

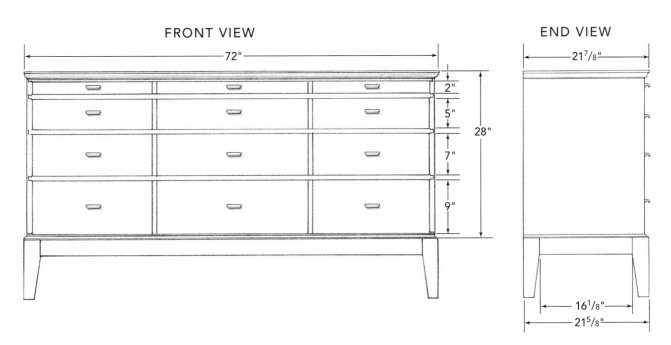

72"

2"

5"

7"

9"

28"

21$\frac{7}{8}$"

16$\frac{1}{8}$"

21$\frac{5}{8}$"

BUILDING THE CHEST STEP-BY-STEP

CUT LIST FOR TRIPLE DRESSER

Case			
1	Top	13/16 in. x 21 in. x 70¼ in.	cherry
1	Bottom	13/16 in. x 20 3/16 in. x 70¼ in.	cherry (low grade)
2	Sides	13/16 in. x 21 in. x 28 in.	cherry
8	Web frame rails	13/16 in. x 2½ in. x 68⅞ in.	maple
8	Runners	13/16 in. x 2½ in. x 15 7/16 in.	maple
8	Runners	13/16 in. x 3½ in. x 15 7/16 in.	maple
8	Drawer guides	13/16 in. x 1½ in. x 18 7/16 in.	maple
4	Face frame rails	13/16 in. x 13/16 in. x 69⅝ in.	cherry
2	Face frame stiles	13/16 in. x 13/16 in. x 26⅝ in.	cherry
4	Support frame rails	13/16 in. x 13/16 in. x 68⅞ in.	maple
2	Support frame stiles	13/16 in. x 13/16 in. x 26⅝ in.	maple
2	Back end stiles	13/16 in. x 2½ in. x 27 9/16 in.	cherry
2	Back intermediate stiles	13/16 in. x 2½ in. x 24 9/16 in.*	cherry
2	Back rails	13/16 in. x 2½ in. x 66⅜ in.*	cherry
3	Back panels	13/16 in. x 20½ in. x 23 5/16 in.	cherry
2	Base rails	1¾ in. x 2¾ in. x 68 in.†	cherry
2	Base end rails	2¾ in. x 2¾ in. x 18⅛ in.†	cherry
4	Legs	2¾ in. x 2¾ in. x 10 in.	cherry
Moldings			
1	Front base molding	½ in. x ½ in. x 71¼ in.	cherry
2	End base moldings	½ in. x ½ in. x 21½ in.	cherry
1	Front top molding	⅞ in. x 1¼ in. x 72 in.	cherry
2	End top moldings	⅞ in. x 1¼ in. x 21⅞ in.	cherry
2	Molding keys	5/16 in. x ⅜ in. x 21 in.	cherry

CUT LIST FOR TRIPLE DRESSER

Drawers

2	Drawer fronts	13/16 in. x 2 in. x 20½ in.	cherry
1	Drawer front	13/16 in. x 2 in. x 26 in.	cherry
6	Drawer sides	⅝ in. x 2 in. x 20 in.	maple
2	Drawer backs	⅝ in. x 1¼ in. x 20½ in.	maple
1	Drawer back	⅝ in. x 1¼ in. x 26 in.	maple
2	Drawer fronts	13/16 in. x 5 in. x 20½ in.	cherry
1	Drawer front	13/16 in. x 5 in. x 26 in.	cherry
6	Drawer sides	⅝ in. x 5 in. x 20 in.	maple
2	Drawer backs	⅝ in. x 4¼ in. x 20½ in.	maple
1	Drawer back	⅝ in. x 4¼ in. x 26 in.	maple
2	Drawer fronts	13/16 in. x 7 in. x 20½ in.	cherry
1	Drawer front	13/16 in. x 7 in. x 26 in.	cherry
6	Drawer sides	⅝ in. x 7 in. x 20 in.	maple
2	Drawer backs	⅝ in. x 6¼ in. x 20½ in.	maple
1	Drawer back	⅝ in. x 6¼ in. x 26 in.	maple
2	Drawer fronts	13/16 in. x 9 in. x 20½ in.	cherry
1	Drawer front	13/16 in. x 9 in. x 26 in.	cherry
6	Drawer sides	⅝ in. x 9 in. x 20 in.	maple
2	Drawer backs	⅝ in. x 8¼ in. x 20½ in.	maple
1	Drawer back	⅝ in. x 8¼ in. x 26 in.	maple
8	Drawer bottoms	¼ in. x 19¾ in. x 19⅝ in.	birch plywood
4	Drawer bottoms	¼ in. x 25¼ in. x 19⅝ in.	birch plywood
12	Drawer pulls	½ in. x 1¼ in. x 42 in.	bird's-eye maple

* Including a ⅜-in. x 1½-in. x 1-in. tenon on each end.
† Including a ½-in. x 2-in. x 1-in. tenon on each end.

YES, IT'S BIG AND COMPLICATED. But the triple dresser is like any other chest of drawers. It consists of a case, a base, some trim, and drawers.

To narrow your focus and make the casework less daunting, I've segmented that part of the work into constructing the case shell, then the web frames, then the face and support frames, and so forth. But all the work begins with selecting, laying out, and dressing the lumber.

PREPARING THE STOCK

Preparing the stock is not a perfunctory task for Seward. He'll typically invest a full day in visiting a familiar, well-stocked hardwood

Photo A: Using chalk and a tape measure, lay out all the parts for the project on the boards. Take your time and endeavor to make the best aesthetic use of the material you have. Make sure, for example, that adjacent drawer fronts are taken from the same board.

lumberyard, consulting with the owner, and picking through and evaluating boards. For this triple dresser, he purchased all the boards taken from a huge log. The board lengths exceeded 12 feet, and some ranged up to 21 in. wide.

Back in his shop, he evaluates his stock in the rough and lays out the parts on the individual boards (see **photo A**). He then crosscuts and rips the parts to rough size, flattens one face, and planes to rough thickness (which is to within ⅟₁₆ in. of the final thickness). The parts are then placed in stickered stacks in his shop until he needs them.

Dimensioning the parts

1. Lay out all the parts on the rough stock using chalk.
2. Crosscut and rip the parts to rough dimensions, which means to about 2 in. over final length. Label each part on the end of the board.
3. Joint a face and an edge of each piece.
4. For each piece, plane the second face flat and parallel to the first, reducing the thickness to ⅟₁₆ in. over final dimension.

5. Rip each piece to final width plus ³⁄₁₆ in.
6. Stack the pieces with stickers until needed. When you are ready to use a part, joint and plane it to the final dimensions.

MAKING THE CASE

The shell of the case is formed by the top, the bottom, and two sides. Seward had wide, long boards of attractively figured curly cherry to work with, and he was able to take both sides and the top from a single board. You probably will have to glue up stock for these parts.

A very narrow strip of the bottom is visible in the finished piece. If the bottom is made from a secondary wood, you could apply a cherry edge band to it. Seward simply used a lower-quality cherry board. The case bottom is ¹³⁄₁₆ in. narrower than the sides and top. The ¹³⁄₁₆-in.-thick back assembly will be rabbeted into the top and sides, but overlap the bottom. The width difference must be accounted for in laying out the dovetails (see "Case Dovetail Layout" on p. 146).

Dadoing the sides and bottom

1. Before actually making the dadoes for the web frames, evaluate the sides and determine which face of each is best. Mark this as the outside.

2. Lay out the dadoes on the sides, as shown in "Case Side Layout."

3. Make the cuts on the table saw, using a dado head and crosscut box. (Seward has an after-market sliding table on his cabinet saw, which makes this relatively easy to do accurately.) Use the rip fence to locate the dado. (The dadoes can also be cut with a router, guided along a clamped-on straightedge.) Size the dadoes for a "sliding fit." When you mount the web frames to the case, you will slide them into the dadoes from the front after the case itself is glued up. While you don't want the web frames to be loose, you don't want them to bind and seize during assembly.

4. Cut each individual dado in both sides before repositioning the fence for the next one, so the dadoes will align across the case.

Dadoing the bottom

Two dadoes are needed in the bottom, primarily for the drawer guides but also to house the bottom ends of the face frames. Make these cuts before dovetailing the case parts.

Cutting the dovetails

Hand-cut dovetails always have variations, so you must mark the pin board and tailboard of each joint before you begin to lay out and cut any of them. Also mark the best face of the top, so you orient it properly as you cut the dovetails. Finally, mark a reference face on the bottom, so it can be oriented consistently.

There are lots of ways to cut dovetails. This is the approach Seward uses.

1. Clamp a tail board (either case top or case bottom) on the benchtop, with its end overhanging the edge.

2. Pencil the layout of the tails on the stock (see "Case Dovetail Layout" on p. 146). Scribe the base of the tails with a layout knife to cut

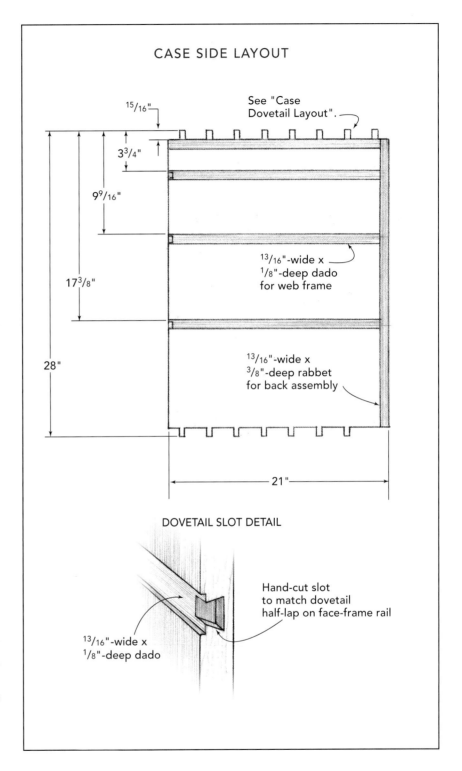

CASE SIDE LAYOUT

See "Case Dovetail Layout".

$^{15}/_{16}$"

$3^3/_4$"

$9^9/_{16}$"

$17^3/_8$"

28"

$^{13}/_{16}$"-wide x $^1/_8$"-deep dado for web frame

$^{13}/_{16}$"-wide x $^3/_8$"-deep rabbet for back assembly

21"

DOVETAIL SLOT DETAIL

Hand-cut slot to match dovetail half-lap on face-frame rail

$^{13}/_{16}$"-wide x $^1/_8$"-deep dado

Photo B: Remove the waste between the tails quickly with a jigsaw. Cut in from the board's edge along both layout lines; then nibble to the baseline. Pare the base clean with a chisel.

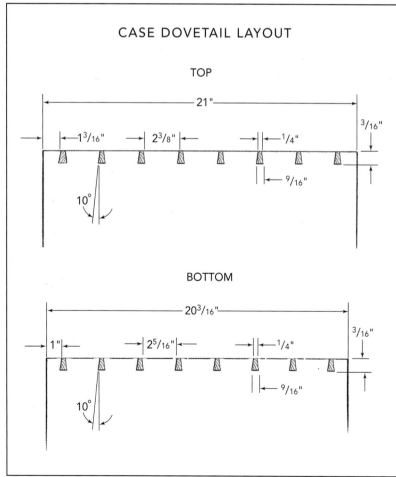

CASE DOVETAIL LAYOUT

TOP

21"

1³/₁₆" 2³/₈" ¹/₄"

³/₁₆"

⁹/₁₆"

10°

BOTTOM

20³/₁₆"

1" 2⁵/₁₆" ¹/₄"

³/₁₆"

⁹/₁₆"

10°

through the surface of the wood and prevent splintering.

3. With a jigsaw, waste the slots between the tails (see **photo B**).

4. Clean up the cut edges with a chisel.

5. Cut the tails on both ends of the case top and the case bottom in this way.

6. Mark out the pins next. Clamp the case parts together, butting the pin boards against the undersides of the tail boards. Make sure everything is properly aligned.

7. Use a layout knife to scribe along the sides of the tails, marking the end grain of the pin board (see **photo C**).

8. Clamp the pin board vertically to the front of the workbench.

9. With a router and a ⅜-in. spiral upcut bit—and working freehand—carefully waste the material from between the pins. Take small bites and gradually plunge down so you maintain control of the router as you work. Seward made a work holder for cutting pins, which looks like a dovetail jig without a template. A stop on it prevents the router from cutting into the holder itself (see **photo D**).

10. Pare the sides of the pins as necessary with a chisel to achieve a perfect fit (see **photo E**).

Photo C: Align the side beneath the top, just the way they will go together. Use a sharp layout knife to slice the side's end grain along the tails.

Photo D: Carefully rout out the bulk of the waste between the pins. Cut as close to the layout marks as you dare. Then pare to the marks with a chisel.

Photo E: A well-fitted joint closes with hand pressure, perhaps supplemented with light taps. If the joint binds, isolate the trouble spots and pare additional material from the pins. If the joint binds dry, it will seize with glue.

Rabbeting the top and sides

The back assembly is housed in $^{13}/_{16}$-in.-wide by $^{3}/_{8}$-in.-deep rabbets cut into the case sides and top; the back overlays the case bottom. The rabbets in the sides are through cuts. In the top, the rabbet is stopped at both ends. You can cut these rabbets with a router, straight bit, and edge guide.

Tip: Use mover's blankets between the sawhorses and the case to prevent accidental damage like scratches and dents.

1. Clearly mark the extent of the rabbet in the top panel.
2. Set up your router with a 1-in.-diameter straight bit and the edge guide. Adjust the guide so the maximum cut will be $^{13}/_{16}$ in. wide. Set the cutting depth to about $^{1}/_{8}$ in.
3. Make a pass on each of the sides and from mark to mark on the top. When routing the top, remember that it's better to stop shy of the marks than it is to cut past them inadvertently.
4. Increase the cutting depth to $^{1}/_{4}$ in. Make a second pass on each part.
5. Increase the cutting depth again, to the full $^{3}/_{8}$ in. Then make a third and final pass on each part.
6. Square the ends of the rabbet in the top with a chisel.

Dry-assembling the case

Several purposes are served by assembling the case now. First, you ensure that the case will go together and that, when assembled, it is square. Second, you enable yourself to make the web frames, the face frame, and the support frame to fit.

1. Lay the case bottom across a pair of sawhorses.
2. Remember not to use glue. Join the sides to the bottom.
3. Set the case top in position, align the pins under the slots, and work the top down into position.
4. Check the case with a square and measure the diagonals to be sure that the unit is square.
5. Leave the case dry-assembled this way while you move ahead and build the web, face, and support frames.

MAKING THE WEB FRAMES

The web frames are made of maple. The front and back rails extend from side to side and join the sides in shallow ($^{1}/_{8}$-in.-deep) dadoes. The runners are mortised into the rails. Their tenons are glued into mortises in the front rail but are left unglued in the back rail.

There is a web frame installed under the top, to serve as a kicker for the top drawers. (So you must make four all together.) The bottom drawers ride on the case bottom.

Cutting the parts to final size

1. Take measurements for the web frames directly from the dry-assembled case. You will be sure the frames fit properly if you do this.
2. If necessary, thickness the stock for the rails and runners to fit the dadoes already cut in the case sides.
3. Crosscut the rails $^{1}/_{4}$ in. longer than interior length of case (or the same as the measurement from dado bottom to dado bottom).
4. Crosscut the runners. Each runner has a 1-in.-long tenon on each end, but there's a $^{1}/_{8}$-in. gap ($^{3}/_{16}$ in. if this is a summertime construction project) between the shoulder of the tenon and the edge of the back rail (so the case side can expand and contract without splitting). Thus cut the runners $1^{7}/_{8}$ in. longer than the distance from rail to rail.

Cutting the joinery

The web frames are assembled with mortise and tenon joints (see "Web Frame Construction"). Seward's practice is to cut the tenons and then rout mortises to accommodate them. He squares the ends of the routed mortises with a chisel rather than rounding or chamfering the tenons. The tenons are $^{3}/_{8}$ in. thick, with a shoulder of a consistent width all around.

1. Set up the table saw with a miter gauge and stop to cut the tenon shoulders $^{7}/_{32}$ in. deep on all four sides of the workpieces.

WEB FRAME CONSTRUCTION

The drawer guides are glued to the runner and front rail but not the back rail. The runner tenons are glued into the front rail mortises; the tenons are left unglued in the back rail mortises.

FRAMES IN CASE

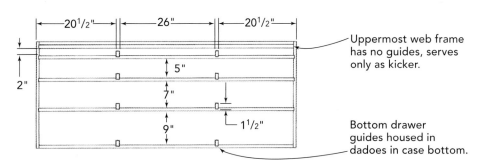

Uppermost web frame has no guides, serves only as kicker.

Bottom drawer guides housed in dadoes in case bottom.

PLAN VIEW

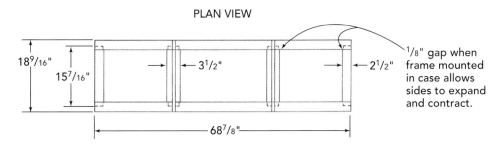

$^1/_8$" gap when frame mounted in case allows sides to expand and contract.

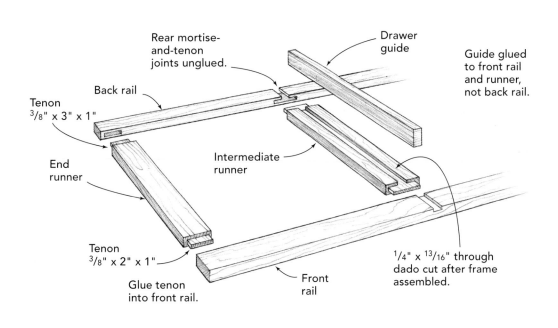

Rear mortise-and-tenon joints unglued.

Drawer guide

Back rail

Guide glued to front rail and runner, not back rail.

Tenon $^3/_8$" x 3" x 1"

End runner

Intermediate runner

Tenon $^3/_8$" x 2" x 1"

Glue tenon into front rail.

Front rail

$^1/_4$" x $^{13}/_{16}$" through dado cut after frame assembled.

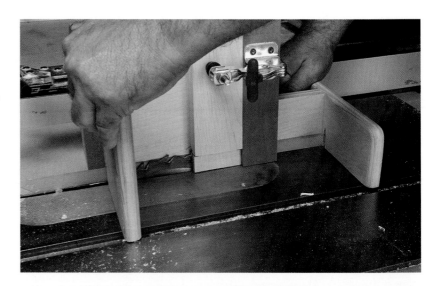

Photo F: Tenons can be sawed quickly and accurately on the table saw. After sawing the shoulders, use a tenoning jig to hold the workpiece while cutting the cheeks.

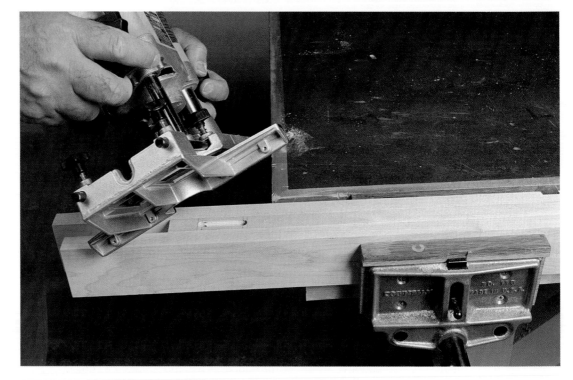

Photo G: Establish the ends of mortises by drilling holes on the drill press. Then clamp two or three rails together in a bench vise and use a plunge router to remove the waste between the bored holes.

2. Use a tenon jig to cut the cheeks of the tenon, leaving each ⅜ in. thick and with an even shoulder all around (see **photo F**).
3. Now rout mortises in the rails with a plunge router. To support the router, stack the rails together, offsetting them, so extended support is offered, even when doing a mortise right at the end of a rail (see **photo G**).
4. Square the corners of the mortises with a chisel and fit each tenon in place.

Assembling the web frames

1. Glue the runners into the mortises in the front rail.
2. Fit the back rail onto the runners' rear tenons, but don't use glue.
3. Apply a clamp to the assembly at each runner. Put two of the clamps across the top surface and two across the bottom.
4. Measure the diagonals to ensure that the assembly is square.

Photo H: Apply glue to the first third of the drawer guide when gluing it into its dado in the web frame. Align its end flush with the front edge of the rail.

Mounting the drawer guides

The case sides double as drawer guides, but guides must be fitted to the web frames at the intermediate runners to complete the drawer-guide system. In this case, the guide is a strip of the secondary wood set into a dado cut into the top surface of three of the web frames (the top frame serves only as a kicker; it has no drawer guides, thus it needs no dadoes). Though the dado is through, the guide is glued to the runner and front rail only.

1. Set up the table saw for dadoing. Install the dado head, and space the cutters to produce a $^{13}/_{16}$-in.-wide cut.

2. Use a cutoff box (or if your saw has it, the sliding table) to hold the web frame, so you get accurate cuts. Set up a stop block, either on the cutoff box fence or the saw's rip fence, to position the near edge of each cut 20⅝ in. from the end of the frame.

3. Set the depth of cut to ¼ in. (If you use a cutoff box, you need to account for the thickness of its bottom in setting the cutter height.)

4. Cut a dado at one end of a frame; then turn it around and cut a second one in the same surface. Make sure the back rail stays in place, since it isn't glued.

5. Trim and dress the guides as necessary to fit the dadoes. The guides, like the runners, should be ⅛ in. shorter than the distance from the face frame to the back, to allow the case to expand and contract.

6. Glue the guides into the dado in the front rail and in the runners (see **photo H**). Make sure each is flush with the front edge of the frame.

7. Glue the guides into the dadoes in the case bottom. Make sure these guides are set back $^{13}/_{16}$ in. from the bottom's front edge—the thickness of the face frame.

Tip: If your diagonal measurements are unequal, apply a bar or pipe clamp along the longer of the diagonals and slowly tighten it. Monitor the effect on the other diagonal. When the difference in the original measurements is split, the assembly should be square.

ASSEMBLING THE CASE

Glue up the case and web frames next. Having the case solidly joined will make it easier to get tight face frame joints. You probably will benefit from a dry run. Follow the sequence below, but don't use glue. Get the clamps set and the necessary cauls in hand. And determine what the problems are without risking a botched glue-up.

The web frame rails are glued into the dadoes, but the runners are not. If you glue a runner to the case side, it'll prevent the side from expanding and contracting; and in dry conditions, the side will crack. You also want to establish the gap between the runner shoulders and the back rail. If you fail to have the correct gap, contraction of the side in dry conditions will jam the runner against the back rail, which could cause the joinery to fail.

As you locate the rails, you need to accommodate the face and support frames, which have yet to be made. Be sure to set the rails back $^{13}/_{16}$ in. from both the front edge of the case and the shoulder of the rabbets in the back. For the face frame, it's okay if the setback is a trifle under, because you can plane it flush with the case edges. But in the back, it is essential to hit the mark. Otherwise, the back assembly won't fit properly.

Gluing up the case

Following a successful dry run, break open the glue and repeat the process for real.

1. Set the case bottom across a pair of sawhorses. Set out the case sides on the workbench. Have the case top and any clamps and cauls you will use at the ready as well.
2. Apply a thin, even coat of glue to the edges of the case bottom tails and the bottom pins on the sides.
3. Join the sides to the case bottom. Make sure the joints are closed tight.
4. Apply glue to the pins on the top of the sides and to the tails on the case top.

5. Set the top in place and force the joints to close tightly.
6. Check the case carefully to ensure that it is square. Check the corners with a square and measure the diagonals, front and back.
7. Apply clamps extending from top to bottom on the front and back of the case. As you do, recheck for square, to ensure that you don't distort the alignment with the clamps.

Installing the web frames

Allow the glue to set; then install the web frames.

1. Cut a scrap or two of the face frame stock to use as gauges in installing the web frames. The face frame (and the support frame in the back) will be $^{13}/_{16}$ in. and must be flush with the case's front edges. You want to be sure you position the web frames properly to avoid fitting problems later on.
2. Slide the front section (the front rail and runners, which are glued together) of a web frame into its dadoes. Because the frame is captured between the face faces, there's no need to glue it in place.
3. Push the frame into the case. Hold the face frame gauge against the front edge of the rail, right at the dado, and tap the frame in. Set it the thickness of the gauge from the front edge of the case. Duplicate the setting at the other end of the frame.
4. Apply glue to the dado at the now-exposed tenons on the ends of the runners.
5. Slide the back rail into the case from the back. As you do this, you must align the tenons on the runners with the mortises in the rail. Remember that you do not glue these joints. Use the gauges, just the way you did at the front, to set the position of the rail. If you made the web frames accurately, the back rail will be inset from the shoulder of the back rabbets by the thickness of the gauge, and you will still have a ⅛-in. gap between the rail and the shoulders of each runner's tenon.

The web frames should not need to be clamped.

MAKING THE FACE AND SUPPORT FRAMES

The face frame consists of square strips of the primary stock, assembled into a grid and edge-glued to the web frames. Where horizontal and vertical strips cross, a half-lap joint is used. The ends of the strips are joined to the casework with a dovetail half-lap joint (see "Face and Support Frame Construction").

The support frame is almost a duplicate, made of the secondary wood. It is mounted in the back of the case to support the web frames and prevent them from sagging under the weight of the drawers. The rails for this frame do not have the dovetail half-laps.

Cutting the parts

With the case assembled, it is easy to cut the parts to fit.

1. Measure the case from the base of one dado to the base of the opposite one.
2. Crosscut the horizontal pieces for the support frame to that length.
3. To determine the length of the face frame rails, add ¾ in. to the dimension measured in step 1. Cut the face frame rails to that length.
4. Measure the case from the base of the drawer guide dadoes in the bottom up to the top.
5. Crosscut the vertical pieces for both frames to that length.

Making the cross-lap joints

1. Mount the dado head in the table saw, adjusting it to make a ¹³⁄₁₆-in.-wide cut. Adjust the cutting height to ¹³⁄₃₂ in., half the thickness of the frame stock. (Because of the way the cross-laps are used, if you cut a tad deep, no one but you will ever know. In the assembled piece, it will not be evident.)
2. Lay out the cross-laps on all the stiles. You can do this by setting the stiles in place in the dry-assembled case and marking them where they cross the web frames.
3. Now cut the cross-laps in the stiles on the table saw.

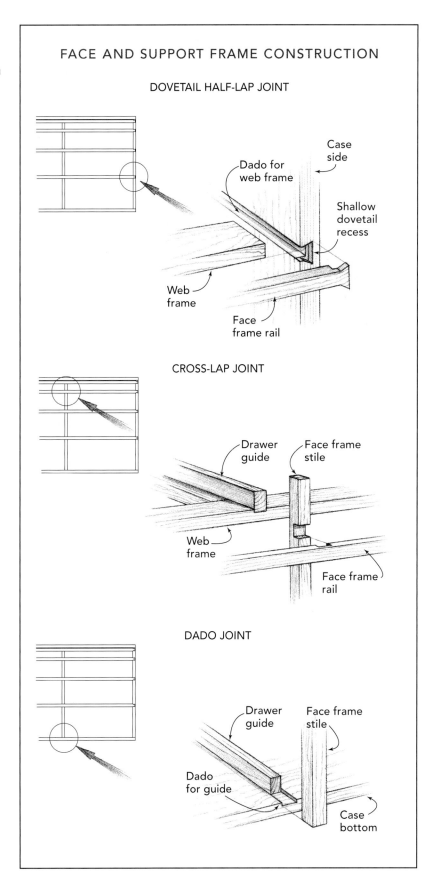

FACE AND SUPPORT FRAME CONSTRUCTION

DOVETAIL HALF-LAP JOINT

Case side
Dado for web frame
Shallow dovetail recess
Web frame
Face frame rail

CROSS-LAP JOINT

Drawer guide
Face frame stile
Web frame
Face frame rail

DADO JOINT

Drawer guide
Face frame stile
Dado for guide
Case bottom

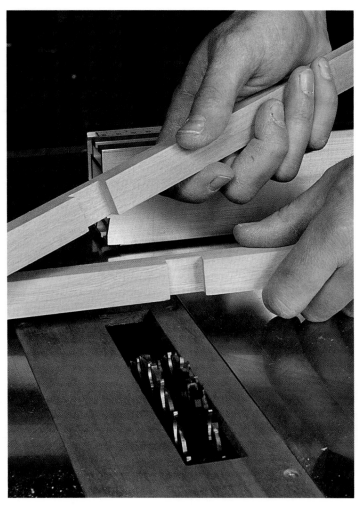

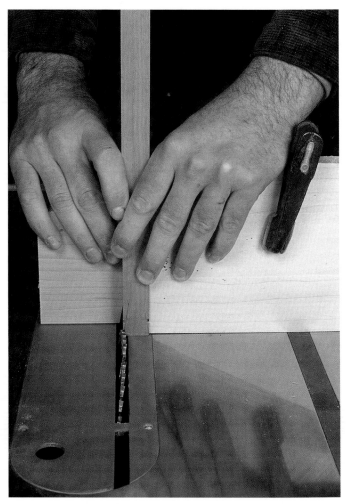

Photo I: Make test cuts on scraps of the working stock to set up the dado head width and the cut depth. The width of the cross-laps must be precise to achieve a good appearance. The exact depth is less critical, so long as the cuts are not too shallow.

Photo J: Cut the angled edges that form the half-lap dovetail on the table saw. The specific angle is less critical than aligning the shoulders of the angled cuts with that of the half-lap.

4. Lay out the cross-laps on the support frame rails. Set the stiles in place and mark them where they cross through the laps in the stiles.
5. Cut these cross-laps on the table saw (see **photo I**).
6. Lay out the cross-laps on the face frame rails. Again, set the stiles in place; then place the rails. With the face frame, the rails must be centered so an equal amount overlaps the case side on either end. Mark where the rails intersect the stiles and where they overlap the sides.
7. Cut these cross-laps.

Making the dovetails

1. Reinstall the regular sawblade in the table saw. Tilt the blade to 10 degrees. Adjust the height.
2. Attach a wooden facing with a stop to the miter gauge. Stand a face frame rail on end, brace it against the stop and miter gauge facing (which should be high enough to lend appropriate support), and feed it past the blade to cut one side of the tail.
3. Turn the rail around and cut the second side (see **photo J**).

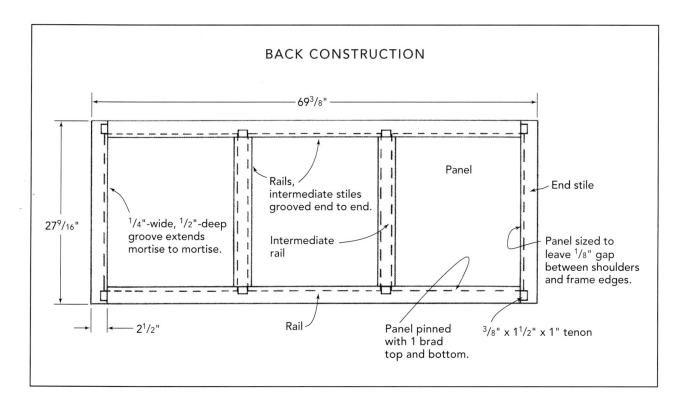

BACK CONSTRUCTION

69³/₈"

27⁹/₁₆"

2¹/₂"

¹/₄"-wide, ¹/₂"-deep groove extends mortise to mortise.

Rails, intermediate stiles grooved end to end.

Intermediate rail

Panel

End stile

Panel sized to leave ¹/₈" gap between shoulders and frame edges.

Rail

Panel pinned with 1 brad top and bottom.

³/₈" x 1¹/₂" x 1" tenon

Cutting the dovetail slot

1. Use the tails cut on the rail ends as templates to lay out the slots in the case sides. Set the rails in place on the case. Scribe around the tails with a layout knife.

2. Waste the slots with a router and ¹/₄-in. spiral upcut bit; then clean the slot to the line with a chisel.

MAKING THE BACK

The back is a frame-and-panel assembly, with all parts taken from ¹³/₁₆-in. stock. The assembly is designed to have a ¹/₈-in. gap between the shoulders of the panels and the edges of the rails and stiles (see "Back Construction"). Make sure, as you construct the assembly, that you adjust part dimensions as necessary to maintain the gap.

Cutting the frame parts

1. Measure the case to confirm the dimensions of the parts.

2. Trim the back stiles and back rails as necessary, based on the actual dimensions of your chest.

Cutting the frame joinery

For Seward, the joinery of the back assembly begins with the tenons. When using mortise-and-tenon joints, he cuts the tenons first, then routs mortises to accommodate them. With this frame-and-panel unit, he next routs the panel groove.

1. Cut ³/₈-in.-thick by 1-in.-long tenons on the back rails and intermediate stiles. Make the shoulders the same as those on the web frame runners and cut them the same way too.

2. Rout mortises in the rails and the back end stiles for the tenons. Size the mortises for the tenons, aligning them on the center of the stock's edge.

3. Cut the panel grooves to complete the frame joinery. Seward makes them ¹/₄ in. wide and ¹/₂ in. deep, centered across the stock. Rout them from mortise to mortise, not through and through. The end stiles and the rails are grooved on only one edge, but the intermediate stiles are grooved on both edges.

Making the back panels

1. Cut the three back panels to the dimensions specified on the cut list. It's a good idea to check the measurement against the chest and the frame before actually cutting the wood.

2. Form a ¼-in.-thick by ½-in.-long tongue around the perimeter of each panel. This can be done by routing or sawing a rabbet around all four edges on both sides of each panel.

Assembling the back

1. Plane some scrap stock down to a ⅛-in. thickness. Cut the stock into 2-in. to 3-in.-long spacers. You need at least 12.

2. Apply glue to the mortises with a brush.

3. Join the intermediate stiles to one of the rails.

4. Fit the panels into the grooves in these parts. Use no glue, of course. Fit a spacer into the gap between the panel's shoulders and each rail or stile.

5. Fit the second rail in place. As you close the joints, insert spacers between this rail and the panel shoulders.

6. Add the end stiles. Again, insert spacers.

7. Clamp the assembly and measure the diagonals to ensure that it is square. Make sure the assembly is flat.

8. From the back of the assembly (the side that will face the inside of the case), shoot brads into the frames and through the panel tongues. Each panel gets just two brads. One is located in the top tongue, centered from side to side, and the second is in the bottom tongue, also centered. Doing this will allow the panels to expand and contract, but will hold them in position in the frame.

9. Fit the assembled back into the case to check its fit. Plane the edges as necessary to achieve a good fit. Don't mount the back just yet. Set it aside until after the drawers have been constructed.

MAKING THE BASE

The base is a leg-and-apron assembly. It's a separate element, and the case is not attached to it (though you could fasten the assemblies together). The case is plenty heavy and isn't going to shift. The molding is attached to the base, not to the case.

Cutting the parts

The legs are made from 12/4 stock, the rails from 8/4 stock, and the molding from 4/4 stock. All the parts were prepared earlier and should be ready to go.

1. Plane the legs and rails to remove that final ¹⁄₁₆ in., reducing the legs to a 2¾-in. thickness and width and the rails to a 1¾-in. thickness.

2. Plane the molding stock to ⅝ in. thick, but leave the stock wide. You will rout the profile and then rip the molding strips from it.

3. Measure the case to confirm that the lengths specified in the cut list will produce a base that will fit it. Crosscut the parts to the final dimensions.

Cutting the joinery

The legs and rails are joined with mortise-and-bare-faced-tenon joints. The tenons are ½ in. thick and 1 in. long and offset, so the front cheek is flush with the face of the rail. The top and bottom shoulders are ⅜ in. wide (see "Base Construction").

1. Cut the tenons on both ends of each rail. Use whatever method you employed to cut the tenons in the runners and the back frame members.

2. Cut the mating mortises in the legs. (Square the corners of routed mortises with a chisel.)

Taper the legs

The two adjoining inside faces of each leg (those with the mortises) taper from the edge of the rail to the foot. The two outside faces are plumb.

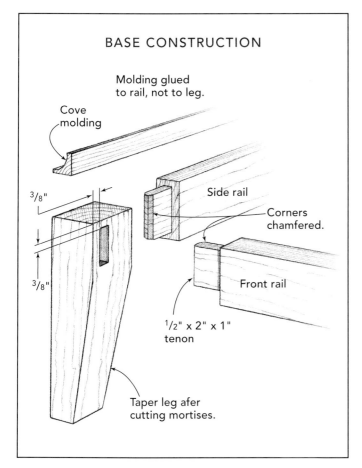

BASE CONSTRUCTION

Molding glued
to rail, not to leg.

Cove
molding

Side rail

Corners
chamfered.

$3/8"$

$3/8"$

Front rail

$1/2" \times 2" \times 1"$
tenon

Taper leg afer
cutting mortises.

Photo K: Taper the
two inner surfaces of
the legs below the
rail. Because of the
stock thickness, this
is a bandsaw job.

1. Lay out the tapers on each leg.
2. Cut the tapers. You can do this on the bandsaw using masking tape to hold the parts together after the first cut (see **photo K**).
3. Sand the cut surfaces smooth and flat.

Joining the legs and rails

1. Dry-assemble the legs and rails to check the fit of the joints and to set up clamps and cauls.
2. Apply glue with a brush to the insides of the mortises in the legs and to the tenons. A thorough, even application is better than a liberal but spotty one.
3. Insert the tenons in the mortises, joining the rails to the legs, and forming the base. Apply clamps, placing cauls between the work and the clamp jaws to prevent the clamps from crushing or otherwise marring the wood.

4. Make sure the assembly is square and that the base's top surfaces are flat and true. You don't want it to be twisted.

It is common to approach the assembly of a unit like the base in stages. You join each pair of legs to a rail (either front/back or side), and clamp them. After the glue sets, you combine the two leg-and-rail subassemblies with the remaining two rails, forming the base. This approach is sound, but it protracts the process.

MAKING
THE MOLDINGS

Making the base molding

1. The base molding profile is a ½-in.-radius cove. Cut it on the router table or with a molding head in the table saw.

2. Set the case on the base and align it carefully. The back edges of the case should be set back ⅛ in. from the back of the base, and you should have a ⅝-in.-wide ledge along the front and across both ends. The molding will be attached to this surface of the base.

3. Measure the front of the case and cut the front molding strip. Miter the molding's ends.

4. Glue the molding to the base (not the case). You do want the molding tight to the case. The grain of the molding parallels that of the rail. Only where the molding crosses the tops of the legs do you have a cross-grain situation. You can avoid gluing the molding to the leg tops or you can anticipate that this short span of long-grain to end-grain glue-surface won't cause problems.

5. Cut two over-long end molding strips for the sides. Miter the appropriate ends. Apply glue to the mitered ends of the molding, and to the bottom of the molding. Bond the moldings to the base, tight against the case.

6. After the glue sets, trim the ends of the end moldings flush with the back of the base and the case.

Making the top molding

The molding around the top of the case is cut with a large ogee router bit. It is glued to the case along the front, but it shouldn't be glued to the sides. There, it is mounted on several dovetail keys, so the side can expand and contract. The ogee profile is big, but it can be cut easily on the router table with the right bit.

1. Start with three top molding blanks ⅞ in. by 1¼ in. and crosscut each several inches longer than the cut list specifies.

2. Sketch the desired profile on the end of one of the blanks to aid you in adjusting the bit heights and fence positions.

3. Chuck the ogee bit in the router and adjust the bit height against the sketch on the end of the blank. Adjust the fence so you make a shallow initial cut. (The full profile is cut in three passes, so you re-adjust the fence position after each cut.)

4. Rout the profile. Make the first cut on all three molding strips. Re-adjust the fence position and rout again. For the final pass, set the fence using the sketched profile. Rout all three strips.

Applying the front top molding

Because the grain direction of the molding is the same as the case top, it can simply be edge-glued in place.

1. Miter one end of the molding.

2. Clamp the molding to the case front, with the miter properly aligned at the corner.

3. Mark the other end of the molding.

4. Unclamp the strip and then miter the second end.

5. Glue and clamp the molding to the case front. Be sure the ends are properly aligned, so you can get very tight miter joints between this molding and the strips to be mounted across the case sides.

Routing the slots in the side moldings

1. On the router table, cut a ¼-in.-wide by ⁵⁄₁₆-in.-deep groove with a straight bit in the back of the two side molding strips. Use a featherboard.

2. Without moving the router-table fence, switch to a ⅜-in. dovetail bit, and re-rout the groove, turning it into a dovetail slot.

Cutting the dovetail keys

Use the same bit and height setting in the router table to make the dovetail keys. Just move the fence.

1. Take the stock that you planed earlier to a ⅝-in. thickness. You need a 3-in.-wide by 24-in.-long strip to make the keys and a 5-in. to 6-in. strip to use for setting up the router-table fence.

2. Set the fence so just the tip of the bit is exposed. The goal is to cut a dovetail that will just fit the slot that's already been routed in the moldings. Cut the setup piece twice, forming a tail along the edge.

Tip: Cherry is easily burned when routing. To reduce—if not eliminate—this problem, use a very sharp bit and a fast feed rate.

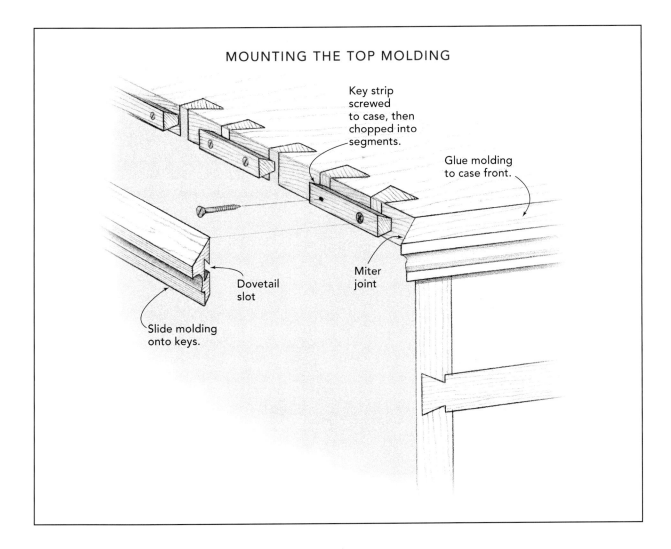

MOUNTING THE TOP MOLDING

Key strip screwed to case, then chopped into segments.

Glue molding to case front.

Dovetail slot

Miter joint

Slide molding onto keys.

3. Check the fit of the dovetail in the slot. If necessary, re-adjust the fence and recut the tail. You want a snug fit, but you don't want the tail to bind.

4. Rout each edge of the workpiece, forming the dovetail key.

5. Rip the dovetail key from the workpiece on the table saw.

6. Plane the cut edge (the narrower edge) to reduce the thickness of the key strips. Having the keys a bit thinner than the height of the slots will pull the molding tight to the case.

Mounting the dovetail keys

The keys are screwed to the case side in a strip, then chopped or sawed into pieces (see "Mounting the Top Molding"). Leaving the key as an unbroken strip would have the same bad effect on the case side as gluing the molding to it would. But aligning individual pieces would be next to impossible. Hence the approach shown here.

1. Use a piece of molding to position the key properly. Hold the molding against the side, flush with the top. Mark the top and bottom of the dovetail slot at the front and back of the case, using a knife. Guide the knife along a straightedge to connect the marks.

2. Lay out the individual keys on the key strip. Divide the strip into six keys, with a ½-in. gap between them. Mark screw locations, about ½ in. to either side of a key's center. Saw about halfway through the strip at the division lines.

Photo L: Rather than trying to align individual keys, divide a strip into keys by sawing halfway through it. Screw the strip to the case side, kerfs in. Saw in to the kerfs to remove the waste.

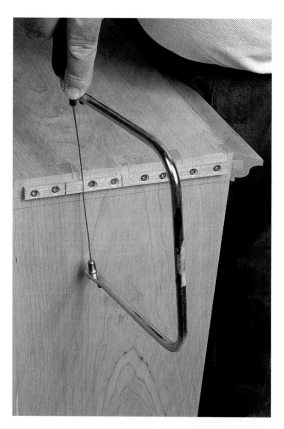

3. Apply a dab of glue at the center point of each key, and apply the strip to the case side, on the position lines. Place the kerfed face of the strip against the case. Use a few ¾-in. brads to temporarily secure the strip.

4. Drill and countersink pilot holes for #4 by ¾-in. screws at the appropriate spots; then drive the screws. Pull the brads.

5. Carefully saw the strip into pieces, removing the waste between the individual keys (see **photo L**).

Mounting the side molding

1. Test the fit of the molding. It should slide onto the keys without binding. If the fit is too tight, a little sanding of the tops of the keys may be necessary.

2. Miter the front end of the molding.

3. Put a dab of glue on the miter and on the first 2 in. of the case. Tap the molding home.

4. Clamp the molding across the case at the miter.

5. After the glue has dried, saw off the excess at the back of the case.

MAKING THE DRAWERS

Making the drawers is a critical and time-consuming step, because the drawers have to be well made and tightly fitted to the case. Moreover, there are 12 drawers in eight different sizes to make. Two keys to success are to work methodically and to clearly identify the parts. The fundamental design is the same for all the drawers (see "Drawer Dovetail Layout").

Cutting the parts

At the initial stock-prep stage, Seward cuts the drawer fronts and backs so they are ⅛ in. or more wider than the drawer openings. The parts are similarly oversize in terms of stock width. The initial task in drawer building is to trim down the parts but still leave them tight in the openings. As you work, judge part sizes by fitting them in (or against) the actual openings in the case.

1. Tailor the heights of the fronts and sides to the openings. Measure the height of the openings; then rip the parts to fit. Remember that the taller the opening, the more of an expansion gap is needed. The 2-in.-high top-drawer fronts and sides can fit very snugly, but the 9-in.-high bottom-drawer fronts and sides need as much as an ⅛-in. expansion gap.

2. Measure the case depth, subtract ³⁄₁₆ in., and crosscut all the sides to exactly that length. Use a stop on the table-saw crosscut box or use the power miter saw fence to ensure this.

3. Trim the length of the fronts and backs, leaving them about ¹⁄₁₆ in. longer than the opening widths.

4. Label all the parts as they are cut and fitted.

Grooving the fronts and sides

1. Measure the thickness of the plywood you will use for the drawer bottoms. Plywood is notorious for being undersize and for varying in thickness from sheet to sheet and even within a sheet. Match the groove width to the plywood thickness.

DRAWER DOVETAIL LAYOUT

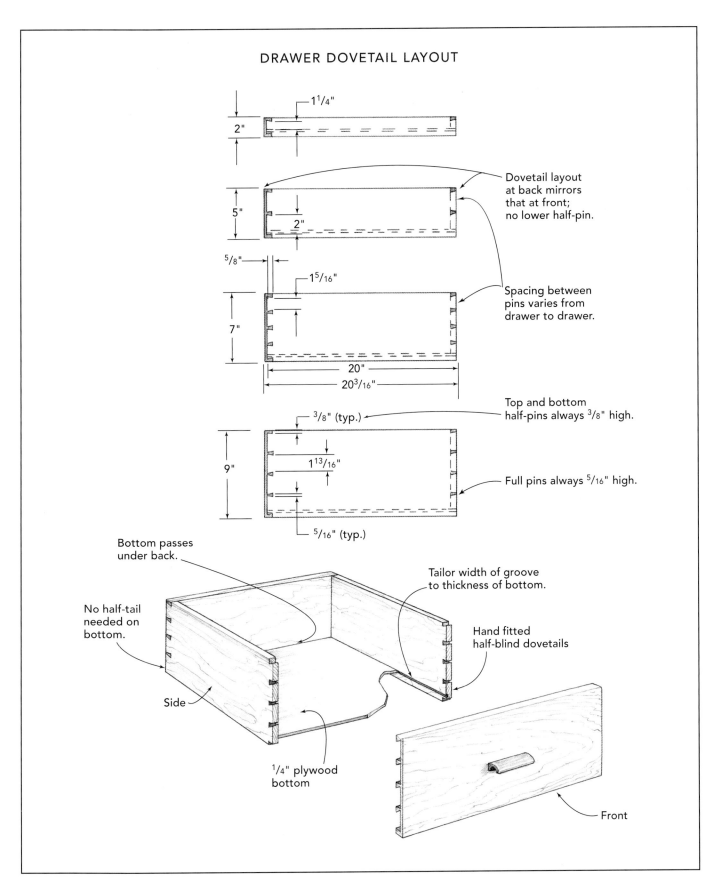

1¼"

2"

Dovetail layout
at back mirrors
that at front;
no lower half-pin.

5"

2"

Spacing between
pins varies from
drawer to drawer.

⁵⁄₈"

1⁵⁄₁₆"

7"

20"

20³⁄₁₆"

Top and bottom
half-pins always ³⁄₈" high.

³⁄₈" (typ.)

1¹³⁄₁₆"

9"

Full pins always ⁵⁄₁₆" high.

⁵⁄₁₆" (typ.)

Bottom passes
under back.

Tailor width of groove
to thickness of bottom.

No half-tail
needed on
bottom.

Hand fitted
half-blind dovetails

Side

¼" plywood
bottom

Front

Photo M: Butt the drawer front to the underside of the drawer side and slip a scrap of the plywood into the grooves to align them perfectly. Then use a marking knife to scribe the edges of the tails onto the end grain of the drawer sides.

2. Groove the sides and front for the bottom. Locate the upper edge of the groove ¾ in. from the bottom edge of the part. Size the groove to match the plywood. The grooves can be cut quickly on the table saw. (Be sure, when you set the fence, to include the blade width in the ¾-in. measurement.)

Note: The uppermost drawers in the original chest have jeweler's cloth bonded to the drawer bottoms as the only concession to their intended use. Seward bonded the cloth to the bottoms before they were installed, so he had to make the grooves in these drawers wider than the others. He determined what that width would have to be at this stage and cut those grooves extra wide.

Cutting the dovetails

When laying out and cutting dovetail joinery on drawers, Seward has basic guidelines he follows. On this dresser, the sides and backs are ⅝ in. thick and the fronts, ¹³⁄₁₆ in. thick. The same dovetail can be laid out and cut at the back of the drawer and the front. At the back, it is through and at the front it is half-blind.

Seward cuts a ⅜-in.-wide half-pin at the top and the bottom. All the other pins are ⁵⁄₁₆ in. wide. He spaces these pins evenly between the two half-pins and typically has wide tails. He will vary the number of pins, depending on the height of the drawer. He uses a 10-degree angle for his dovetails (see "Drawer Dovetail Layout" on p 161).

The joinery work progresses on a joint-by-joint basis, meaning that you lay out and cut both tails and pins for a joint before moving to the next joint. The drill is to lay out and cut the tails on the sides; then transfer the tail outlines to the pin boards (the drawer fronts and backs). Cut the pins and then test-fit the mating parts together. If the fit is right, move to the next joint. Every joint has to be marked individually, of course. Seward uses chalk to label the mating parts so there are no mix-ups.

1. Clamp the side to the benchtop.
2. Lay out the tails on the sides.
3. Cut out the waste between the tails the same way you did for the case dovetails.
4. Clamp the pin board (either the drawer front or back) under the side with its end aligned to scribe lines on the inside of the sides, front, and back. Use a scrap of the bottom plywood fitted to the grooves to align the side and the front.
5. Then use a knife to mark the edges of the tails on the end grain of the pin board (see **photo M**).
6. With a router and straight bit, and working largely freehand, waste the material from between the pins. Set the cutting depth of the router to ²¹⁄₃₂ in. (that is, ⅝ in. plus ¹⁄₃₂ in.), so

that the pins will be slightly proud of the side's surface when the joint is assembled. Rout shy of the layout marks.

Seward's work holder for cutting pins has an adjustable router stop on top. He can set it to hold the ³⁄₁₆-in. setback from the face of the drawer.

7. With a chisel, pare to the layout marks.

8. Assemble the joint. You should be able to push or lightly tap a dovetail joint closed. Any tighter a fit, and the joint will bind when glue is applied.

Fitting the drawers

1. Join the front and sides of the drawer, without glue, to test the fit. Slide this unit into the appropriate opening in the chest case. Though the front is still too long to fit into the opening, you should be able to determine if the unit is fitting properly. It should slide easily into the opening, right up to the point that the ends of the front fit the face frame. But it shouldn't have side play; just a snug fit.

2. Repeat the process with the sides and back. Fit the unit into the case from the back.

3. If a drawer binds, you need to pare the slots between the pins deeper. If it is loose, you have to decide whether it is so loose that you need to make a new front or back.

Assembling the drawers

1. Joint just the bottom edge of the fronts, removing ¹⁄₃₂ in.

2. Glue up the front, sides, and back. Clamp only briefly. Check that the assembly is square and flat. Set it on saw table or jointer bed; use a try square or measure the diagonals.

3. Sand the pins flush after the glue has set. Seward has a fixture he clamps to the benchtop. The drawer is set on it so he can belt-sand the pins (see **photo N**). Check the fit of each drawer as you go.

4. With a rabbet plane, plane the bottom edge of each drawer front. This is a part of fitting each drawer in the case visually. The goal is to have a consistent gap on three sides of each drawer front—sides and bottom. (The gap at the top varies from tier to tier; here the gap

Photo N: After the drawer is assembled, but before the bottom is installed, sand the pins flush with the side. A scrap of plywood with notches for the drawer front and back, clamped to the workbench, holds the work.

has to allow for expansion of the wood.) Seward tries to match the gap on either side with one beneath the drawer front. The drawer rides on its sides, not its front.

5. Rout a groove for the drawer pull. Use the same fixture to hold the drawer, front up. Fit a plunge router with an edge guide to locate the grooves in relation to the top edge. Because the pulls are well below eye-level, they tend to look lower than they really are. To compensate, Seward aligned them ¼ in. above center (the bottom edge of the pull is right on the horizontal centerline of the drawer, in other words).

DRAWER PULL PROFILE

TOP

FRONT

2¹/₄"

3"

BACK

¹/₄"

2¹/₂"

SIDE

³/₈"   ⁷/₈"

³/₄" r.

¹/₄"   ¹/₄" r.

SEQUENCE OF CUTS FOR DRAWER PULLS

1. Mill a strip of stock to the correct thickness and width.
2. Form the tongue by cutting rabbets on the table saw.
3. Round over the top surface with a ¾-in.-radius roundover bit in the table-mounted router.
4. Rout a finger groove on the underside with a ½-in.-diameter corebox bit in the table-mounted router.

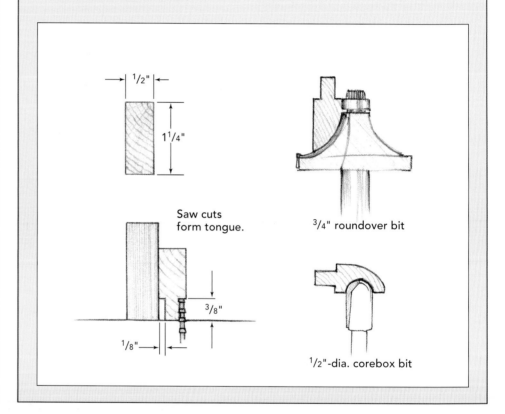

¹/₂"

1¹/₄"

Saw cuts
form tongue.

³/₄" roundover bit

³/₈"

¹/₈"

¹/₂"-dia. corebox bit

Making the drawer pulls

Each drawer has a pull mounted almost exactly in the center. The pulls are made from bird's-eye maple. On the center bank of drawers, which are slightly wider than the outer drawers, the pulls are 3 in. wide. On the outer drawers, they are 2½ in. wide.

1. Mill a strip or two of bird's-eye maple to ½ in. by 1¼ in. You need at least 36 in. of the stock to make all the pulls.

2. Cut the profile on the strip (see "Sequence of Cuts for Drawer Pulls").

3. Crosscut the strip into 3-in. lengths for the center drawers and 2½-in. lengths for the outer drawers.

4. Miter the ends of each pull.

5. Trim ¼ in. from each end of the tongue, forming a stub tenon. Then round off one corner and fit the pull into the grooves (see **photo O**).

Completing the drawers

1. Insert the bottom. Slide it into the grooves and seat it in the bottom of the groove in the drawer front. Secure it with two or three screws through the bottom into the edge of the back.

2. Fit a pull into the slot mortise on the drawer front. Secure it with a pair of screws driven into it from inside the drawer.

FINISHING UP

The only tasks left are to install the case back and to apply a finish. Seward finishes his case-work inside and out, so having the back off makes that a little easier to accomplish.

Like many furniture makers, Seward has his own finish recipe. Typically, he'll apply five or six coats. As the finish builds up, less and less of it will soak into the wood.

The first coat is oil, straight out of the can. He uses Minwax Antique Oil. Subsequent coats will be a blend of equal parts of the oil and a varnish. He uses either Behlen's Rock-Hard Tabletop Varnish or Behlen's Water-White Restoration Varnish. Between coats, Seward buffs out the finish with #0000 steel wool.

The drawers get the same finish, though not as many coats. Wax is applied to the drawer sides and runners.

The final construction task, when the finish is completed, is to install the back. Seward fits the back in place and secures it with brads fired in with a power brad driver. This makes it possible to remove the back, should that ever be necessary.

QUEEN ANNE CHEST ON FRAME

The Queen Anne style was introduced around 1725 and initially remained in vogue for only about 30 years. It gave way to the busier Chippendale style, which yielded to the Federal styles, which were overpowered in the nineteenth century by the Empire and Victorian styles. Nearly 300 years later, though, Queen Anne is still popular, still manufactured, and still reproduced.

One craftsman making a living reproducing furniture in the Queen Anne style (along with other eighteenth-century styles) is Glen Huey. Huey works with his father, doing business as Malcolm L. Huey & Sons of Middletown, Ohio. His design for this chest grew out of a familiarity with the character of the style, rather than a particular piece. It displays many of the style's markers, including the graceful cabriole legs and the fluid, curving lines of the aprons. The simple lipped drawers with their highly polished brass butterfly pulls and escutcheons are typical of Queen Anne chests. Even the chest-on-frame form is a Queen Anne marker.

The chest-on-frame form covers a lot of territory, from diminutive silverware chests perched on slender-legged stands to gargantuan chests of drawers on squat stands. This chest on frame is, to me, a happy medium. It weds a practical yet attractive chest of drawers with a graceful stand, and both forms are enhanced.

Queen Anne Chest on Frame

TRADITIONAL CONSTRUCTION distinguishes the chest on frame. The primary joinery is all dovetails and mortises and tenons, challenging to execute well but very simple in concept. Some key components are simply nailed together.

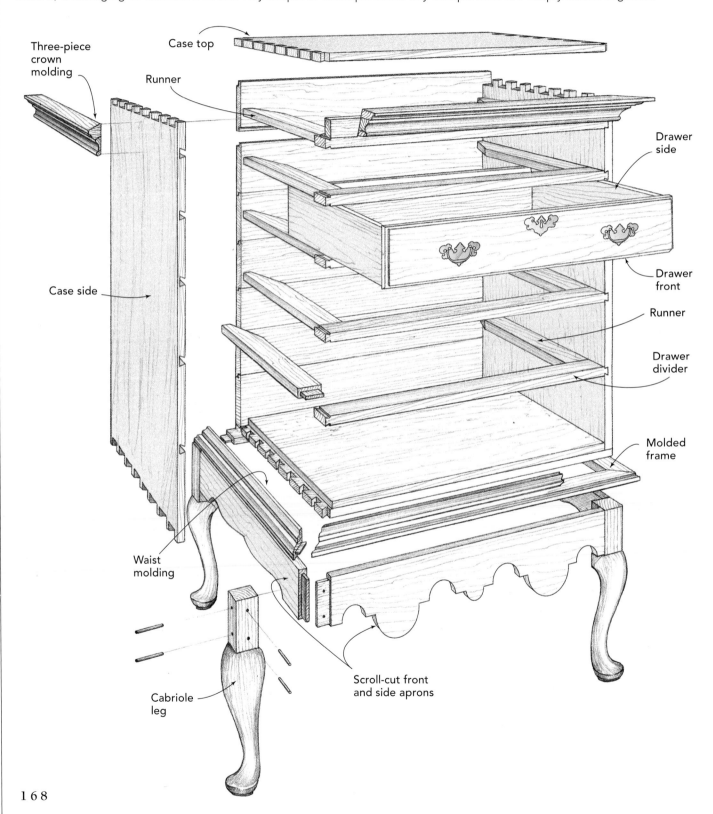

Three-piece crown molding

Case top

Runner

Drawer side

Case side

Drawer front

Runner

Drawer divider

Molded frame

Waist molding

Scroll-cut front and side aprons

Cabriole leg

BASIC CASE, FRONT ELEVATION

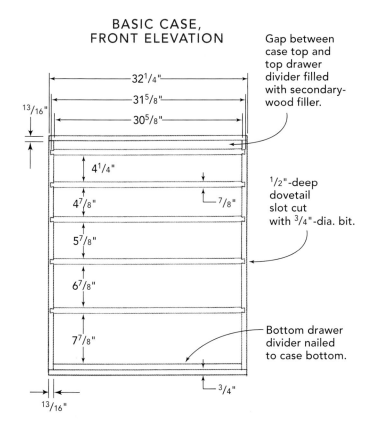

32 1/4"
31 5/8"
30 5/8"

13/16"

4 1/4"
4 7/8"
7/8"
5 7/8"
6 7/8"
7 7/8"

13/16"

3/4"

Gap between case top and top drawer divider filled with secondary-wood filler.

1/2"-deep dovetail slot cut with 3/4"-dia. bit.

Bottom drawer divider nailed to case bottom.

SECTION VIEW

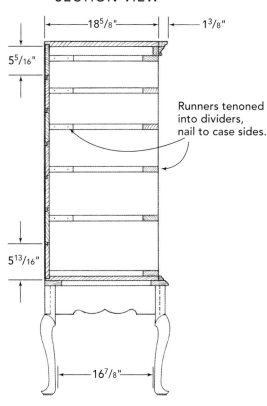

18 5/8"
1 3/8"

5 5/16"

5 13/16"

16 7/8"

Runners tenoned into dividers, nail to case sides.

FRONT VIEW

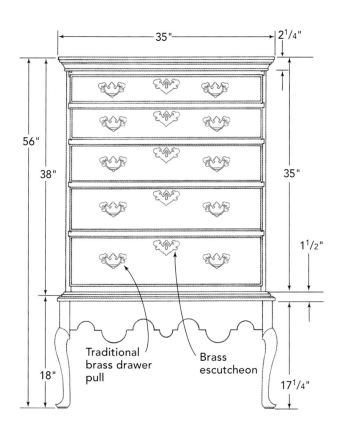

35"
2 1/4"

56"
38"
35"

1 1/2"

18"
17 1/4"

Traditional brass drawer pull

Brass escutcheon

BACK VIEW

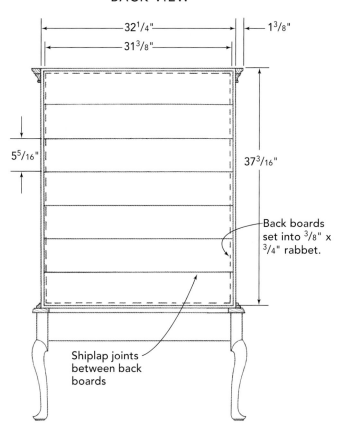

32 1/4"
31 3/8"
1 3/8"

5 5/16"

37 3/16"

Back boards set into 3/8" x 3/4" rabbet.

Shiplap joints between back boards

BUILDING THE CHEST STEP-BY-STEP

CUT LIST FOR QUEEN ANNE CHEST ON FRAME

Stand			
4	Cabriole Legs	2¾ in. x 2¾ in. x 17¼ in.	maple
1	Front apron	⅞ in. x 6½ in. x 32⅞ in.	maple
2	Side aprons	⅞ in. x 4½ in. x 18 3/16 in.	maple
1	Back apron	⅞ in. x 4½ in. x 32⅞ in.	poplar
1	Font molded frame	¾ in. x 2¾ in. x 35¼ in.	maple
2	Side molded frames	¾ in. x 2¾ in. x 20⅛ in.	maple
1	Back molded frame	¾ in. x 2¾ in. x 31¾ in.	poplar
2	Splines	¼ in. x 1½ in. x 3¼ in.	poplar
Case			
2	Case sides	13/16 in. x 18⅝ in. x 38 in.	maple
1	Case top	13/16 in. x 18⅝ in. x 32¼ in.	maple
1	Case bottom	¾ in. x 18⅝ in. x 32¼ in.	poplar
5	Drawer dividers	⅞ in. x 2½ in. x 31⅝ in.	maple
10	Runners	⅞ in. x 2½ in. x 16 in.	poplar
1	Bottom drawer divider	⅞ in. x 2½ in. x 30⅝ in.	maple
2	Runners	⅞ in. x 2½ in. x 15⅜ in.	poplar
1	Filler	⅞ in. x 1 7/16 in. x 30⅝ in.	poplar
1	Back board	⅝ in. x 5 5/16 in. x 31⅜ in.	poplar
6	Back boards	⅝ in. x 5 13/16 in. x 31⅜ in.	poplar
Moldings			
1	Front waist molding	¾ in. x ¾ in. x 33¾ in.	maple
2	Side waist moldings	¾ in. x ¾ in. x 19⅜ in.	maple
1	Front crown cove molding	⅞ in. x 1⅜ in. x 34 in.	maple
2	Side crown cove moldings	⅞ in. x 1⅜ in. x 19½ in.	maple
1	Front crown astragal	¼ in. x ⅜ in. x 33 3/16 in.	maple
2	Side crown astragals	¼ in. x ⅜ in. x 19⅛ in.	maple
1	Front crown cap	⅞ in. x 1⅜ in. x 35 in.	maple
2	Side crown caps	⅞ in. x 1⅜ in. x 20 in.	maple

CUT LIST FOR QUEEN ANNE CHEST ON FRAME

Drawers			
1	Drawer front	13/16 in. x 4½ in. x 31⅛ in.	maple
1	Drawer front	13/16 in. x 5 1/16 in. x 31⅛ in.	maple
1	Drawer front	13/16 in. x 6 1/16 in. x 31⅛ in.	maple
1	Drawer front	13/16 in. x 7 1/16 in. x 31⅛ in.	maple
1	Drawer front	13/16 in. x 8 1/16 in. x 31⅛ in.	maple
2	Drawer sides	½ in. x 4 3/16 in. x 17¾ in.	poplar
2	Drawer sides	½ in. x 4¾ in. x 17¾ in.	poplar
2	Drawer sides	½ in. x 5¾ in. x 17¾ in.	poplar
2	Drawer sides	½ in. x 6¾ in. x 17¾ in.	poplar
2	Drawer sides	½ in. x 7¾ in. x 17¾ in.	poplar
1	Drawer back	½ in. x 3 7/16 in. x 30½ in.	poplar
1	Drawer back	½ in. x 4 in. x 30½ in.	poplar
1	Drawer back	½ in. x 5 in. x 30½ in.	poplar
1	Drawer back	½ in. x 6 in. x 30½ in.	poplar
1	Drawer back	½ in. x 7 in. x 30½ in.	poplar
5	Drawer bottoms	9/16 in. x 17⅝ in. x 30 in.	poplar
Hardware			
5	Chippendale-style brass escutcheons		from Horton Brasses; item #C602se
10	Chippendale-style brass bail-type pulls	2½-in. boring	from Horton Brasses; item #C602S
	Clout nails	1½ in.	from Horton Brasses; item #N7

THE CHEST ON FRAME consists of three major components: the stand, the case, and the drawers. Whether you start with the case or the stand is up to you. Hold off on the drawers until you have the case constructed and can build the drawers to fit it. In any event, the first major undertaking is the selection and preparation of the stock.

Preparing the Stock

Choosing the stock for your chest of drawers is a critical process. The choice of lumber can bump your chest of drawer project from pretty good to fantastic. The truth is that the tiger maple lumber that Huey used to construct his chest on frame has a great deal to do with the piece's impact. Picture it in plainsawn maple—which is the wood used for the construction photos for this project.

The width of the boards used is also significant. Visible seams in glued-up panels detract from the appearance of the chest. A plus with this project is that only two visible panels—the case sides—are really wide.

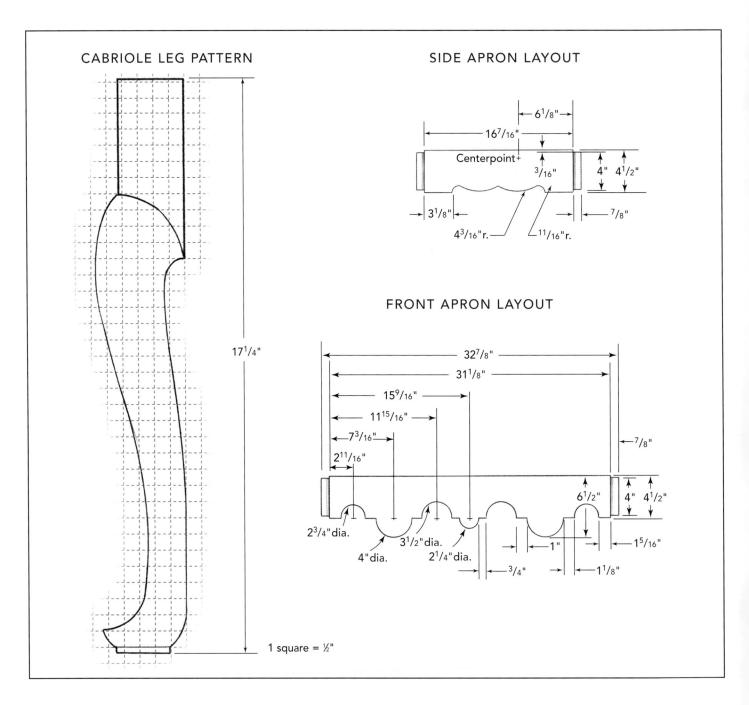

CABRIOLE LEG PATTERN

SIDE APRON LAYOUT

6 1/8"

16 7/16"

Centerpoint

3/16"

4" 4 1/2"

3 1/8"

4 3/16"r.

11/16"r.

7/8"

17 1/4"

FRONT APRON LAYOUT

32 7/8"

31 1/8"

15 9/16"

11 15/16"

7 3/16"

7/8"

2 11/16"

6 1/2"

4" 4 1/2"

2 3/4"dia.

3 1/2"dia.

4"dia.

2 1/4"dia.

1"

1 5/16"

3/4"

1 1/8"

1 square = 1/2"

Cutting the parts to rough size

Beyond the initial choice of lumber, the proper handling and preparation of the wood you do buy are important. Allow the lumber to acclimate to your shop's temperature and humidity. Then lay out all the project parts on the boards. Cut them to rough size, dress the boards, and again allow them to acclimate.

1. Stack the lumber for the project in the shop with stickers between the boards. This allows air to circulate around them. Allow 1 week or more for the lumber to acclimate before cutting it up and dressing the pieces.

2. Lay out the parts on the stock. The project calls for several different stock thicknesses. As you lay out the pieces, allow extra length and

STAND CONSTRUCTION

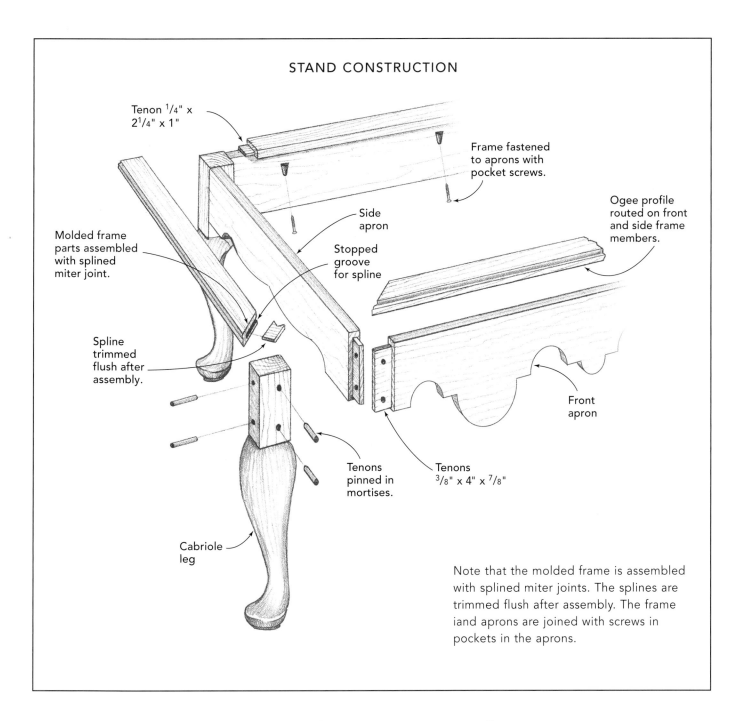

Tenon ¹/₄" x 2¹/₄" x 1"

Frame fastened to aprons with pocket screws.

Ogee profile routed on front and side frame members.

Molded frame parts assembled with splined miter joint.

Side apron

Stopped groove for spline

Spline trimmed flush after assembly.

Front apron

Tenons pinned in mortises.

Tenons ³/₈" x 4" x ⁷/₈"

Cabriole leg

Note that the molded frame is assembled with splined miter joints. The splines are trimmed flush after assembly. The frame iand aprons are joined with screws in pockets in the aprons.

---

width for the parts. Account for everything on the cut list.

3. Crosscut and rip the parts from the rough stock.

4. Joint and plane all the parts of the same thickness at the same time. Stack and sticker the stock.

BUILDING THE STAND

Now tackle the most challenging element of the chest on frame. That's the stand (or frame), with its graceful cabriole legs and its scroll-cut aprons.

Making the leg template

It sounds routine, but making the leg template is critical to the appearance of the chest. The legs won't be well-proportioned and attractive if the template isn't. The template is made of ¼-in. plywood, which is inexpensive. You can make several examples of the template and be judgmental about their contours. Use the one that looks the best.

1. Cut a strip of ¼-in. plywood for the template.
2. Transfer the layout from the cabriole leg pattern to the template blank (see "Cabriole Leg Pattern" on p. 172). You can do this several ways. One is to pencil a ½-in. grid on the template blank. Select points on the pattern grid and plot them on your grid. Another is to enlarge the drawing with a photocopier and use spray adhesive to bond the copy to the template.
3. Saw the template on the bandsaw; then shape and fair the contour with rasps, files, and sandpaper.

Making the leg blanks

The blanks for the legs are milled from 12/4 stock. Avoid gluing up stock to create the blanks; the glue lines will be evident in

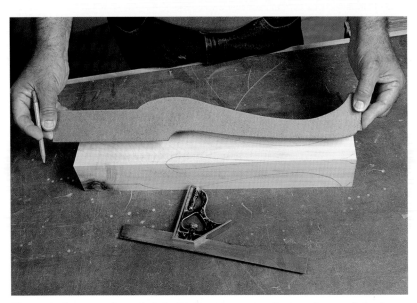

Photo A: Lay out the cabriole leg's curves on adjacent faces of the leg blank using the plywood template. Mark out the post block and foot pad on the blank's ends and locate the centers for turning the pad.

the finished leg and may clash with the graceful curves. For the best results, you want blanks with the grain pattern oriented on the diagonal across the ends and straight grain along each face.

1. Cut the blanks to the dimensions specified on the cut list.
2. Use the leg template to lay out the cutting lines on adjacent faces of the blank (see **photo A**).
3. Mark out the pad dimensions on both ends of the blank. Set a combination square on the template and adjust it to the width of the pad. Then use the square to lay out the pad width on both ends of the blank.
4. Locate the center of the pad on both ends. You'll use these center marks a bit later to turn the pad on the lathe. The center marks indicate the center of the pad, but they won't be the center of the blank.

Cutting the mortises

While the leg blank is completely square, cut the mortises for the aprons. You won't be able to do it easily once the leg is shaped.

1. Lay out the mortises. Huey uses one large tenon to join an apron to a leg.
2. Cut the mortises. No particular technique need be used. You can rout them, cut them with a hollow-chisel mortiser, or chop them by hand.

Cutting the basic leg shape

The curved contours of a cabriole leg are roughly formed with two bandsaw cuts. The foot pad is turned with the leg mounted in the lathe. Then the contour is smoothed and refined with hand tools.

1. Begin by cutting to the lines on one face of the leg. Save the waste as you cut it off.
2. Tape the waste back in place. Use masking tape or packing tape; the type isn't that important.

Photo B: Tape the waste from the initial bandsaw cut back in place and roll the blank to make the second cut. The two cuts give you the leg's basic shape.

3. Turn the leg blank onto the next face and cut to the second set of lines. As you make these cuts, the leg will be resting on that taped-in-place waste (see **photo B**).

Turning the foot pad

1. Mount the roughly shaped blank in the lathe. Use the centers you marked out before bandsawing. Set the tool rest; to ensure that it won't hit the blank, turn the leg by hand.
2. As the lathe spins the leg, use a gouge to begin forming the pad. Turn a cylinder that is the size of the pad at its largest point. Check frequently with calipers as you work to ensure that you don't make the pad too small.
3. Use a pencil to mark the top of the pad. Then cut a shallow line with a skew chisel.
4. Round the pad from the bottom up to the parting line you just cut. Use a gouge for this work.

Shaping the leg

1. Use a spokeshave to clean up the bandsaw marks (see **photo C**). To hold the leg for this work, capture it in a wood handscrew or a K-body clamp and secure the clamp in the bench vise. You can loosen the clamp to rotate the leg as necessary to give you access to all facets of the workpiece.

Photo C: The shaping of each leg is done by hand. A spokeshave cleans the bandsaw marks from the surfaces, and a cabinetmaker's rasp and files soften and blend the edges. A K-body clamp secures the leg while you work.

2. Switch to a rasp to ease the corners. Rotate the leg as necessary to work all the edges.
3. Shape the top of the pad. Use the rasp initially; then switch to a chisel to pare a shoulder along the parting line cut on the lathe.

Photo D: The several arcs, both positive and negative, making up the front apron contour can be accurately cut with a plunge router and trammel. Cut two arcs (or full circles) at each radius before resetting the trammel.

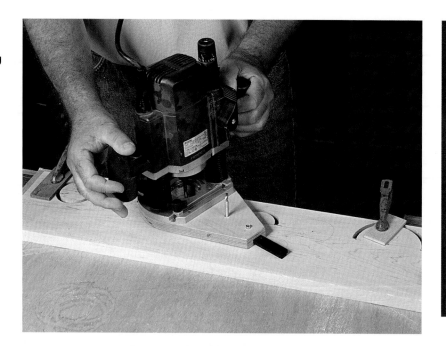

Tip: To achieve good contour on the side aprons, make a template of thin plywood or hardboard. You can saw a contour; then refine it quickly with a rasp or file. You can even start over without remorse if the contour doesn't suit. When you have a shape you like, use it to lay out the apron.

Photo E: Cut the baseline between the convex and concave arcs with a jigsaw. Cut just to free the waste. Clean and square the inside corners with a chisel.

Cutting the tenons on the aprons

With the legs completed, the aprons must be made next. The front and side aprons are contoured on their bottom edges. Depending on your tenon-cutting preference, you may have difficulty cutting the tenons after the aprons are shaped, so make the tenons first.

1. Dress the apron stock to its final thickness and cut the parts to their final dimensions as specified on the cut list.
2. Lay out the tenons as required.
3. Cut the tenons and fit them to the mortises in the legs. The back apron is now complete. The front and side aprons still need to be contoured.

Making the front apron

The front apron's contour is an array of positive and negative half-circles with flats separating them. It is important to the visual balance of the piece that the apron be shaped symmetrically, that the radii of the arcs and drops on one half of the apron match those on the other.

1. Lay out the bottom edge contour on the workpiece, following the layout shown in "Front Apron Layout" on p. 172.
2. Cut the shapes. You can do this on the bandsaw. Or you can use a router and trammel to cut the arcs and a jigsaw to cut the flats (see **photos D** and **E**).
3. Sand or scrape the cut edges as necessary.

Making the side aprons

The side aprons are also contoured, but the line is different.

1. Lay out the contour on the workpieces as shown in "Side Apron Layout" on p. 172.
2. Cut the shapes with a jigsaw or bandsaw.
3. Sand or scrape the cut edges smooth, as necessary.

Assembling the legs and aprons

1. Assemble the legs and the aprons without glue to make sure the joints fit properly. Use the practice run to set up clamps and make sure you have protective cauls on hand.
2. Dismantle the unit and scrape and sand all the parts.
3. Glue up the parts, apply clamps, and set the stand aside while the glue cures.
4. Pin the tenons in the mortises. Use two pins per joint. Locate and drill holes for the pins that penetrate the face of the leg post and the tenon. Whittle the tapered pins from a strip of ¼-in.-square hardwood. Apply a dot of glue to a pin's tip; then drive it home. Pare the pin flush.

Making the molded frame

The stand is topped with a flat frame, joined at the front with splined miter joints and at the back with mortises and tenons. The front and side edges of the frame are molded.

1. Dress the frame parts to their final thickness. Rip the stock to final width, but leave the front and side pieces a couple of inches overlong. Note that the front and side pieces are cut from the maple and the back from the poplar.
2. Rout an ogee profile on the top front edge of the front and side pieces. To make this cut I used a Jesada bit (#659-564; see "Sources" on p. 216) and replaced the standard ½-in. pilot bearing with a ⅜-in. one.
3. Cut miters on both ends of the front frame member and one end of each side frame member.

4. Rout stopped slots in the mitered ends. The idea is to cut through the inner edges of the members but not the outer edges. The slots must be invisible when the frame is assembled. This job is accomplished easily on the router table using a ¼-in. straight bit.
5. Cut the splines for the two miter joints.
6. Assemble the miter joints without glue and establish the position of the rear frame member. Lay out the mortises for it.
7. Cut the mortises for the rear frame member.
8. Tenon the ends of the rear frame member.

Assembling the molded frame

1. Scrape and sand the frame members.
2. Before opening the glue, assemble the frame dry and make sure everything fits properly (see **photo F**). Set your clamps.
3. Glue up the frame and clamp it.

Photo F: The visible frame members have profiled edges and are mitered at the front corners. Hidden splines—almost like loose tenons— strengthen the miters. The poplar back of the molded frame is hidden in the completed piece.

CASE JOINERY

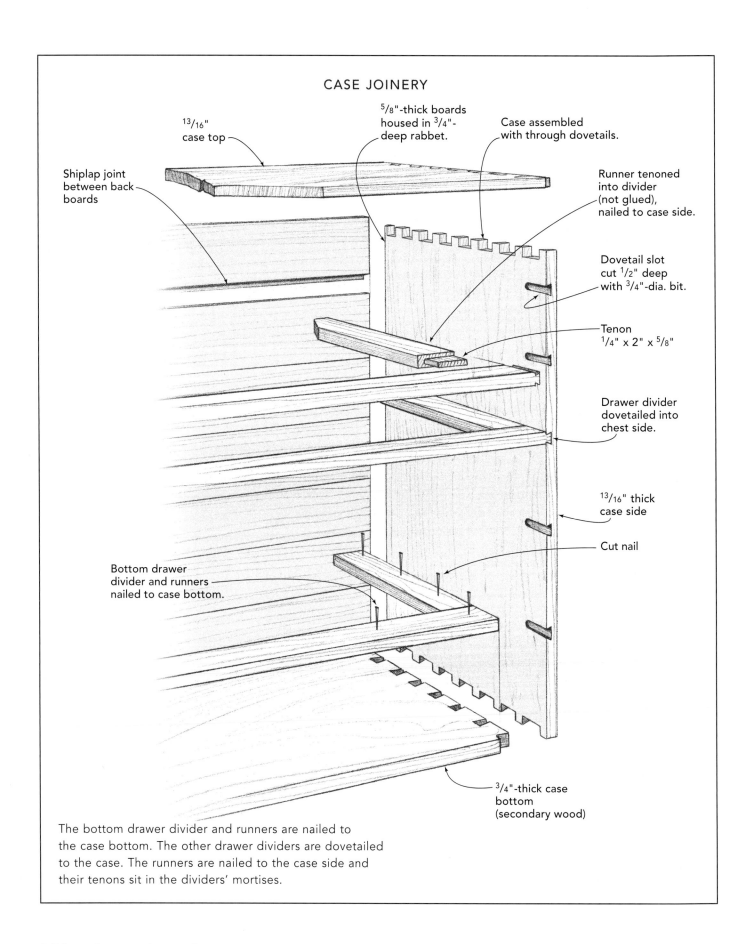

$^{13}/_{16}$" case top

$^5/_8$"-thick boards housed in $^3/_4$"-deep rabbet.

Case assembled with through dovetails.

Shiplap joint between back boards

Runner tenoned into divider (not glued), nailed to case side.

Dovetail slot cut $^1/_2$" deep with $^3/_4$"-dia. bit.

Tenon $^1/_4$" x 2" x $^5/_8$"

Drawer divider dovetailed into chest side.

$^{13}/_{16}$" thick case side

Cut nail

Bottom drawer divider and runners nailed to case bottom.

$^3/_4$"-thick case bottom (secondary wood)

The bottom drawer divider and runners are nailed to the case bottom. The other drawer dividers are dovetailed to the case. The runners are nailed to the case side and their tenons sit in the dividers' mortises.

4. Mount the molded frame on the stand and secure it with pocket screws. Lay the frame face down on the bench and upend the stand on top of it. Center the stand on the frame. Drill the angled pocket and pilot holes in the inside faces of the aprons, and drive the mounting screws.

BUILDING THE CASE

The next major job is building the case. It is nothing more than a big box, assembled with through dovetails. The drawer supports—dividers and runners—are assembled in place.

Preparing the stock

The stock has already been roughly sized and dressed. The initial task is to evaluate the parts to ensure they are still flat and true; then you'll rip, crosscut, and plane them to their final sizes.

1. Mill all the case parts to the desired final thicknesses. You need both primary and secondary woods. Several different thicknesses of stock are used in the case and these are indicated on the cut list.
2. Rip and crosscut the stock to final size. As necessary, glue up narrow boards to produce wide ones.

Cutting the through dovetails

In case construction, the case top (and bottom) usually is the tail board, the sides are the pin boards. With this arrangement, the mechanics of the joint hold the case sides in. Huey cuts all the dovetails by hand, but a dovetail jig works well also.

1. Cut the dovetails using your preferred method.
2. Next dry-assemble the case. The joints should close easily, perhaps with a few taps of a deadblow mallet, and the parts should seat square and tight. Resolve all fitting problems before moving to the next operation.

Rabbeting for the back

The back boards are housed in rabbets in the case sides, top, and bottom. The rabbets are ¾ in. deep, so the back is actually recessed ⅛ in.

1. Cut a stopped rabbet in the back edge of each side. You must stop the cut short of the ends, so it doesn't show when the case is assembled. The rabbet should be ⅜ in. by ¾ in.
2. Cut a stopped rabbet in the back edge of the top and the bottom. The rabbet is the same dimensions as that in each side.

Cutting the sliding dovetail

Eighteenth-century craftsmen typically dove-tailed the drawer dividers into the sides of a case. With practice, a sliding dovetail is only a little more difficult to produce with hand tools than is a dado. But the dovetail provides a mechanical lock that keeps the divider in place with only a touch of glue. And the dividers can then help keep the case sides from bulging outward. With today's power tools, cutting the dovetails is quite simple, and you'll lose none of the benefits.

1. Lay out the locations and extent of the dovetail slots in the sides.
2. Rout the dovetail slots in the case sides. You can produce a very slightly tapered slot using a template, but a slot with parallel edges works fine. Because the slots are stopped, it is advisable to use a double fence guide that traps the router. Otherwise, you may inadvertently widen the slot as you back the router bit out of the cut (see **photo G** on p. 180).
3. Cut a slot or two in a scrap piece of the working stock. Use these in setting up the router table for cutting the tails on the ends of the dividers. You avoid having to fit the test-cut tails to the slots in the case sides themselves if you have this piece available.
4. Set up the router table with the same dovetail bit used to cut the slots. Adjust the extension of the bit above the table and the fence

Photo G: A jig with two fences prevents the router from straying from the desired cut. This is important when cutting the stopped dovetail slots.

position. Dial in the setup, making test cuts on scraps of the working stock, and fitting the dovetails into the setup slots.

5. Rout the dovetails on the ends of the drawer rails.

Making the divider and runner joinery

The drawer runners are cut from the secondary wood, since they don't show. They are tenoned into the drawer dividers and are nailed to the case sides near their back ends. Huey glues the tenon into the mortise. There's enough give by the nail (or by the wood surrounding the nail) that the side can expand and contract without cracking or splitting. This is, in fact, the way eighteenth-century woodworkers constructed such pieces. An alternative approach would be to eschew the glue. As the side expands and contracts, the tenon can move in and out of the mortise.

Photo H: The drawer dividers (often called rails) are dovetailed into the case sides. The dovetail provides a strong mechanical connection, is easy to close with a couple of taps, and doesn't require clamping.

Photo I: The runners, cut from the secondary wood, are tenoned into the drawer dividers (but not glued) and then nailed to the case side. The construction is simple and strong and allows the side to move seasonally.

1. Mortise the drawer dividers. Measure from the shoulder of the dovetail on each divider to lay out the mortise. Figure on making the tenon with a ¼-in.-wide shoulder all around; then size the mortise accordingly. Cut the mortises.

2. Cut the runners to size. In the process, chamfer off the inside back corner. This makes it easier to nail the runner to the case side during assembly.

3. Tenon the opposite end of each runner, fitting the tenons to the mortises.

Assembling the case

1. Glue up the case dovetails, joining the sides, top, and bottom. Check for square by measuring across the diagonals.

2. Install the drawer dividers. One by one, apply a little glue to the dovetails on the drawer dividers and drive them into place (see **photo H**).

3. Nail the bottom drawer rail and runners to the case bottom.

4. Install the runners. Do not glue the runners to the case side. Fit the tenon into the divider mortise; glue it or not, as you prefer. Make sure the runner is square to the face edge of the case side, of course. With a reproduction cut nail, fasten the back of the runner to the case side (see **photo I**).

5. Glue the filler between the top drawer rail and the case top. The grain of this strip of secondary wood runs the same direction of these two case parts, so it can be edge-glued to them. Measure the space, cut a strip of the secondary stock to fit, and glue it in place.

Tip: All too often, nails cause splits. Avoid this disaster by drilling a pilot hole for every nail you use.

CASE MOLDING PROFILES

WAIST MOLDING

Finished profile

STEP 1

├─ 2" ─┤

3/4"

Cut blank.

STEP 2

Reduce bearing size
to get step in profile.

Ogee
bit

Rout ogee profile
on both edges.

STEP 3

Rip moldings to size.

3-PIECE CROWN MOLDING

Finished profile

├─ 1 3/8" ─┤

7/8"

├─ 1" ─┤

Cap

Same ogee
profile used
for waist
molding and
on stand's
molded frame.

├─ 7/8" ─┤

1"

1 sq. = 1/8"

├─ 5/16" ─┤

Cove

3/16" r.

Astragal

TRIMMING OUT THE CHEST

The chest has molding around the top—the crown—and around the transition from case to stand—the waist. To tie the design together, Huey uses the same ogee profile in the crown and for the waist. He also uses it for the molded frame that tops the stand.

The crown is a three-piece assembly, and it is easiest to apply with the case standing on its head on your workbench or assembly stand. So make and apply it first.

Shaping the crown and waist moldings

The crown consists of three moldings, cut separately and then assembled. There is a cove molding, an astragal, and a cap (see **photo J**).

1. Cut the stock for the crown and waist moldings. For each profile of the crown molding, you need a long strip for the front and two short strips for the sides. All should be cut oversize in length initially. You may want to machine the cove and astragal on wide stock and then rip it down. They'll be cut to fit during and after assembly and mounting. The final thickness and width of each molding are given in the cut list.

2. Cut the ogee profile on the crown molding cap on the table-mounted router. Use the bit that cut the molded frame.

3. Using the same bit, rout the waist molding. As in making the other moldings, machine the profile on wide stock; then rip the molding from it.

4. Next, rout the astragal of the crown molding. Use a ⅜-in.-diameter edge-beading bit that produces a step or fillet as part of the profile.

5. Now cut the cove on the crown molding. This is most often done on the table saw, while feeding the stock at an angle across the blade (see "Cove Cutting on the Table Saw" on p. 210 for more information).

Photo J: Machine the crown as three separate profiles: the cove, the astragal, and the cap. Glue them together; then apply them to the case.

Gluing up the crown molding

1. Set the shaped and sized molding strips together to check how they align with one another. You want the profiles to come together virtually seamlessly, so it looks like one piece.

2. Apply glue, join the strips, and clamp them. Clean up any squeeze-out immediately, so you don't have to scrape driplets of hardened glue from the inside corners of the combined profiles later.

Mounting the crown molding

1. Mount the front strip. Cut the molding to fit, mitering the ends. Glue it to the edge of the case and the face of the blocking.

2. Miter the front ends of the molding that extend across the sides. Apply glue to the mitered end and to about 5 in. of the front end of molding. Set the molding strip in place, closing the miter tightly. Nail the back end of the strip to the case side with 1½-in. cut nails. The end can extend past the back of the case for now.

3. Next, trim the moldings flush with the back of the case (see **photo K**).

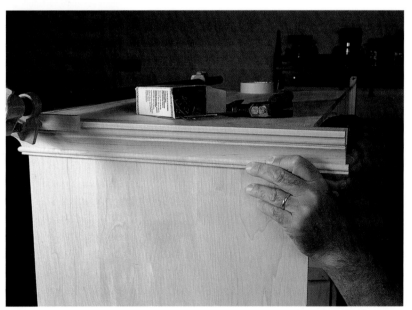

Joining case and stand

1. Set the case on the stand. Align it flush with the back edge of the stand and centered side to side.

2. Nail the case to the stand with 1¼-in. cut nails.

3. Mount the waist molding along the case front. Cut a molding strip to fit, mitering the

Photo K: The best fit is achieved if you focus on the miter joint and let the molding extend beyond the case back. After it is fastened to the case, the molding can easily be sawed flush with the back.

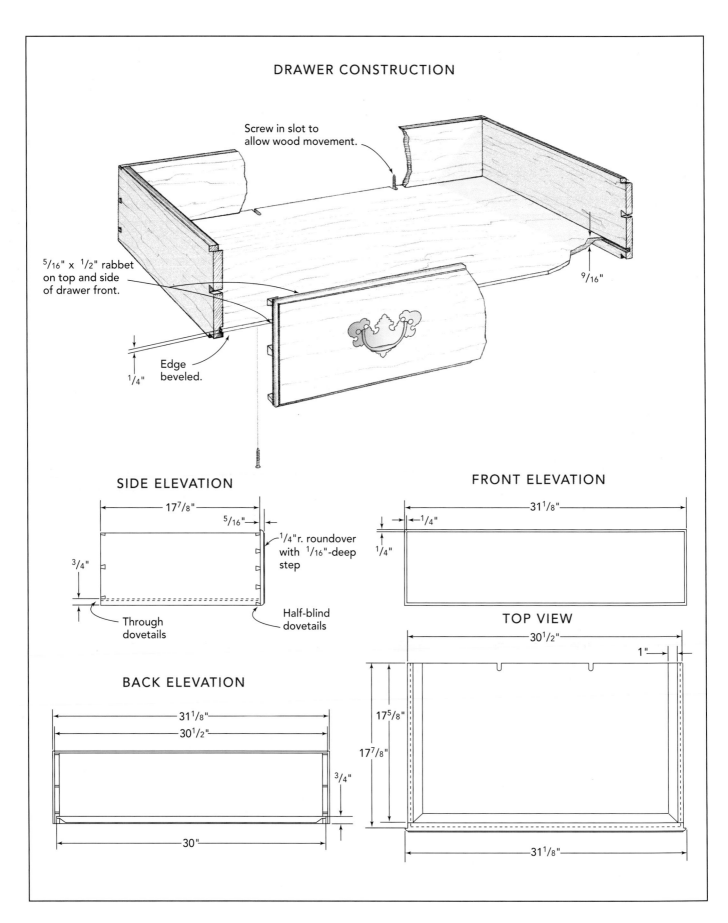

DRAWER CONSTRUCTION

Screw in slot to allow wood movement.

5/16" x 1/2" rabbet on top and side of drawer front.

Edge beveled.

1/4"

9/16"

SIDE ELEVATION

17⁷/₈"

5/16"

1/4"r. roundover with 1/16"-deep step

3/4"

Through dovetails

Half-blind dovetails

FRONT ELEVATION

31¹/₈"

1/4"

1/4"

TOP VIEW

30¹/₂"

1"

17⁵/₈"

17⁷/₈"

31¹/₈"

BACK ELEVATION

31¹/₈"

30¹/₂"

3/4"

30"

ends. Glue it to the edge of the case bottom and to the molded frame.

4. Miter the front ends of the side strips. Apply glue to the mitered end and to about 5 in. of the front end of the molding. Set the molding strip in place, closing the miter tightly. Nail the back end of the strip to the case side. As before, the end can extend past the back of the case.

5. Now trim the side strips flush with the back of the case.

CONSTRUCTING THE DRAWERS

All the drawers are the same, but for their heights. The fronts are made of the primary wood. Each is rabbeted on the ends and top so it sets partly into the drawer opening. All four edges are rounded over, with a $\frac{1}{16}$-in. step defining the profile. The sides are joined at the fronts with half-blind dovetails and to the back with through dovetails. (Huey cuts the dovetails by hand, of course.)

Note that the cut list specifies dimensions for the drawer parts that allow a $\frac{1}{16}$-in. clearance on each side and a $\frac{1}{8}$-in. clearance on the top, assuming your case is dimensioned exactly as shown in the drawings.

Fitting the sides, backs, and fronts

1. Measure the height, width, and depth of each drawer opening.

2. Rip and crosscut the sides to size. Huey keeps the season in mind when fitting the drawers, allowing more clearance in a drawer constructed during the generally dry winter months.

3. Rip and crosscut the fronts to size. Bear in mind that the fronts are rabbeted along the top and across the ends. You want the front's top and ends to overlay the case and drawer divider, but you still need clearance between the case and the shoulders of the rabbets so the drawer doesn't stick.

4. Rabbet the drawer fronts. The cut is $\frac{1}{2}$ in. by $\frac{5}{16}$ in. Choose the best face, mark it, and cut the rabbets into the opposite one. Cut the top and across the ends, but not the bottom. Fit

the fronts to their respective openings. If the front seems too tight, trim the overall size of the front and recut the rabbets to fill the dimension.

5. Finally, rip and crosscut the backs. The backs will be the same length as the front's shoulder-to-shoulder dimension. They will be $\frac{3}{4}$ in. narrower than the sides.

Cutting the joinery

1. Dovetail the sides and backs. These are through dovetails. The top edges are flush and the back is recessed $\frac{1}{4}$ in. so the bottom can slide beneath it. The joints can be laid out and cut by hand or cut with a router and dovetailing jig.

2. Dovetail the fronts and sides. These are half-blind dovetails.

3. Groove the sides and front for the bottom. The groove is $\frac{1}{4}$ in. wide and deep, and the top edge of it is $\frac{3}{4}$ in. from the bottom edge of the parts. The grooves can be cut on the router table or table saw.

Photo L: Drill the mounting holes for the brasses—the drawer pulls and escutcheons—on the drill press before assembling the drawers. A fence clamped to the table aligns all the holes consistently.

Photo M: Bevel the drawer-bottom edges to fit the grooves in the sides and front. To accommodate wood movement, cut two slots through the back edge for clout nails that fasten the bottom to the back.

Drilling for the brasses

Although you are not going to mount the brasses now, this is a good time to lay out the mounting holes for them and to bore the holes on the drill press.

Brasses is a collective name for the bail-type pulls with decorative mounting plates and the matching keyhole escutcheons used on eighteenth- and nineteenth-century furniture. Reproductions of the old brasses are available from a number of sources.

1. Determine where you want the drawer pulls to be on the drawer front. Huey typically centers the plate vertically, but decides how far from each end of the drawer to place them based on what "looks right." Draw a centerline from side to side on one drawer front. Lay out a pair of pulls and find the best side-to-side positioning.
2. Determine where the mounting holes must be located and lay them out on all the drawer fronts.
3. Bore the mounting holes on the drill press (see **photo L** on p. 185).

4. Lay out the location of the escutcheon. Huey places the tip of this plate on the vertical centerline of the drawer front and ⅜ in. down from the edge. Mark this spot on each drawer front and from that point, lay out the mounting hole positions.
5. Drill pilot holes for the nails that attach the escutcheons.

Assembling the drawers

With the fronts rabbeted and all the dovetails cut, the drawers are all but ready for assembly. You need to rout the roundover profile on the fronts. And the bottoms must be beveled to reduce the edge thickness.

1. Rout the profile on the drawer fronts on the router table. Do the end-grain cuts first, and provide good backup so you don't get tearout.
2. Bevel the bottoms. While the bottoms are ⁹⁄₁₆ in. thick, the grooves for them are ¼ in. wide. The edges of the bottom need to be beveled to reduce the edge thickness. It is like raising a panel. You can handplane the edges or rout them with a straight-bevel panel-raising bit. If you do the latter, you'll get a nice

tongue to fit the grooves. You need to bevel both sides but only one long edge.

3. Notch the back edges of the bottoms. To secure a bottom, you drive a fastener or two through it into the drawer back. Because the bottom will expand and contract, some wiggle room around the fasteners is important. The notches, just ⅛ in. or so wide and ½ in. long, provide that. Huey uses 1½-in. clout nails from Horton Brasses to secure the bottoms (see "Sources" on p. 216).

4. Dry-assemble each drawer and fit it to the case. This provides a last chance to rectify fit problems before the drawer is glued up. Because you haven't nailed the case back in place yet, you should have good access to evaluate each drawer's fit at the front and back.

5. Using the glue now, assemble each drawer. Join the sides to the front and back, making sure the assembly is square and rests flat. Slide the bottom into the grooves. The bottom can be glued into the groove in the drawer front, but it doesn't need to be. After drilling pilot holes into the back, drive the nails to secure the bottom (see **photo M**).

FINISHING UP

Only a single task remains before the chest is ready for a finish. That's making and installing the back.

Making the back boards

The back boards are made from the secondary wood. They should be set aside, already dressed, and cut to rough length.

1. Measure the chest and determine the dimensions of the back boards. Bear in mind that the shiplap joinery will take ½ in. away from the effective width of each board.
2. Crosscut the boards to length and rip them to width.
3. Cut the shiplaps. A shiplap is simply a ½-in.-wide rabbet cut to half the thickness of the stock. In this case, that's ⁵⁄₁₆ in. All but two of the back boards are shiplapped on both edges. (The top and bottom boards are shiplapped on one edge only.) But the rabbets are cut into opposite faces, not the same one (see

Photo N). To do the job, you cut a rabbet along one edge; then flip the board over to rabbet the other edge. The cut can be made on the table saw or router table.
4. Nail the boards to the case. Begin at the bottom and work up. Use cut nails and drill a pilot hole for each one.

Applying a finish

Huey stains and seals the entire case, inside and out. The stain adds color and depth to the stock and brings out the figure. He then applies an oil/varnish mix to the outside of the case. The drawer fronts are stained, sealed, and top coated; but the drawer boxes themselves are left unfinished.

When the finish has cured, the escutcheons are nailed to the drawers, and the pulls are mounted.

Photo N: The back is formed of boards shiplapped edge to edge and nailed into a rabbet cut around the inside edge of the case. A shiplap is a rabbet cut to half the thickness of the stock.

TALL CHEST

Country furniture has always appealed to me. It's usually unpretentious, always practical, and frequently quite creative. I also feel a distant kinship to the furniture makers who built it. I picture them as being self-taught, make-do guys, practical and enterprising. So it seems right to me that if I'm going to build a chest of drawers, it should be a traditional country design.

This tall chest is that. It's a vernacular response to the highfalutin highboys, with their imposing size, circumscribed forms, and excessive ornamentation. The tall chest is easier to build and easier to use yet still impressive.

The essential characteristics of the form are few. It has to be tall—typically 5 ft. to 6 ft. high—and it has to have a one-piece case (differentiating it from the chest on frame). It is not typical of any formal style period. For this chest, I used a configuration common during the early nineteenth century from Pennsylvania south through the Shenandoah Valley to North Carolina and even Georgia. I borrowed the frame-and-panel sides and the molding combinations from Chester County, Pennsylvania, chests I've studied.

Like many a nineteenth-century country cabinetmaker, I used a native hardwood—walnut—sawed at a local mill. But as a twenty-first-century guy, I used a lot of power tools and more modern joinery and fasteners to assemble those boards. The result is, I think, unpretentious and practical.

Tall Chest

DESPITE ITS TRADITIONAL LOOK, this chest is constructed with a lot of router-cut joinery. The side frames are assembled with routed cope-and-stick joinery. The sliding dovetails in the casework and the half-blind dovetails in the drawers likewise are router cut.

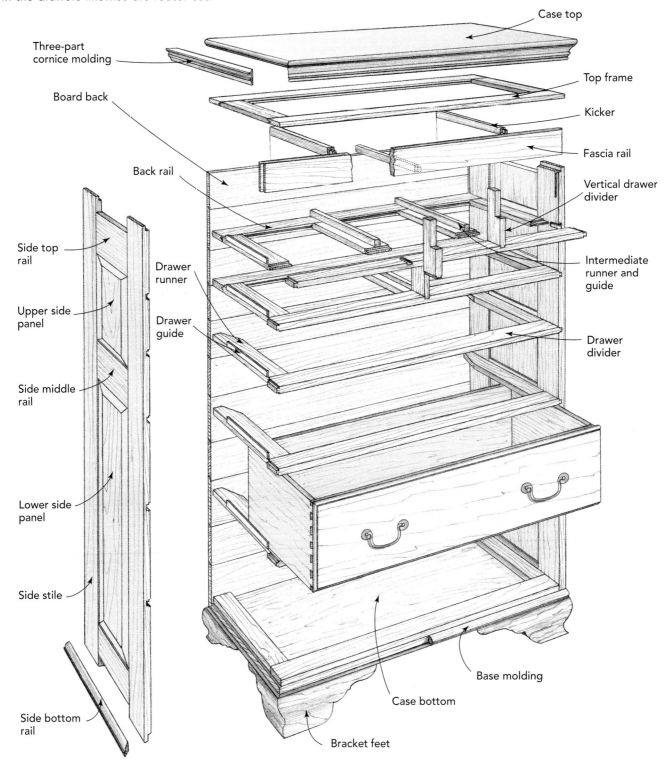

Case top

Three-part cornice molding

Top frame

Kicker

Board back

Fascia rail

Back rail

Vertical drawer divider

Side top rail

Drawer runner

Intermediate runner and guide

Upper side panel

Drawer guide

Drawer divider

Side middle rail

Lower side panel

Side stile

Base molding

Case bottom

Side bottom rail

Bracket feet

FRONT VIEW

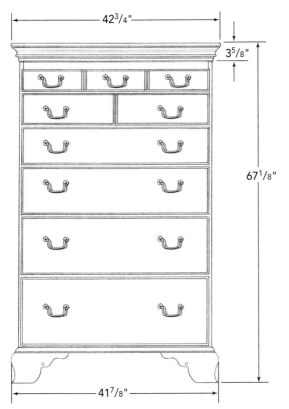

SIDE VIEW

SIDE SECTION

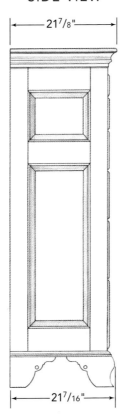

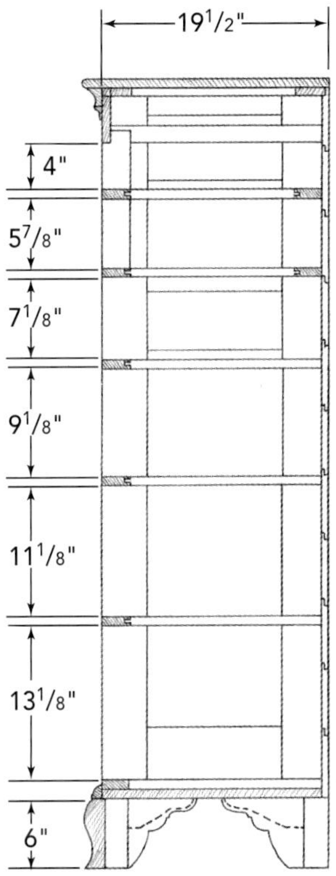

CUT LIST FOR TALL CHEST

Case

4	Side frame stiles	¾ in. x 4 in. x 60⅜ in.	walnut
2	Side frame top rails	¾ in. x 5½ in. x 12¾ in.	walnut
2	Side frame middle rails	¾ in. x 5 in. x 12¾ in.	walnut
2	Side frame bottom rails	¾ in. x 5½ in. x 12¾ in.	walnut
2	Upper side panels	⅝ in. x 12¾ in. x 10¼ in.	walnut
2	Lower side panels	⅝ in. x 12¾ in. x 32¾ in.	walnut
2	Top frame rails	¾ in. x 2½ in. x 38¼ in.	poplar
2	Top frame stiles	¾ in. x 2½ in. x 15½ in.	poplar
1	Bottom drawer rail	¾ in. x 2½ in. x 37½ in.	walnut
2	Bottom drawer runners	¾ in. x 2½ in. x 17 in.	poplar
5	Drawer rails	¾ in. x 2½ in. x 38½ in.	walnut
6	Drawer runners	¾ in. x 2½ in. x 17½ in.	poplar
2	Rear drawer rails	¾ in. x 2½ in. x 38 in.	poplar *(continued on p. 192)*

BUILDING THE CHEST STEP-BY-STEP

CUT LIST FOR TALL CHEST

4	Drawer runners	¾ in. x 2½ in. x 15½ in.	poplar
12	Drawer guides	¼ in. x ⅜ in. x 12 in.	poplar
3	Intermediate drawer runners	¾ in. x 3½ in. x 15½ in.	poplar
3	Intermediate drawer guides	¾ in. x 1 in. x 17 in.	poplar
1	Case bottom	¾ in. x 19½ in. x 38¼ in.	poplar
1	Case top	¾ in. x 21⅞ in. x 42¾ in.	walnut
1	Face rail	¾ in. x 4 in. x 38¼ in.	walnut
3	Kickers	¾ in. x 1½ in. x 19 in.	poplar
1	Vertical drawer divider	¾ in. x 2½ in. x 6⅜ in.	walnut
2	Vertical drawer dividers	¾ in. x 2½ in. x 5¼ in.	walnut
1	Back boards	½ in. x 6 in. x 38¼ in.	poplar
Bracket Feet			
2	Front posts and blocks	2 in. x 6 in. x 9½ in.	poplar
2	Rear posts and blocks	2 in. x 6 in. x 8 in.	poplar
2	Side front blockings	2 in. x 2½ in. x 5 11/16 in.	poplar
2	Side rear blockings	2 in. x 2½ in. x 7 1/16 in.	poplar
2	Front foot facings	1¾ in. x 6 in. x 11⅝ in.	walnut
2	Side rear foot facings	1¾ in. x 6 in. x 9 5/16 in.	walnut
2	Side front foot facings	1¾ in. x 6 in. x 9⅞ in.	walnut
Moldings			
1	Base molding	¾ in. x 1 in. x 40½ in.	walnut
2	Base moldings	¾ in. x 1 in. x 20¾ in.	walnut
1	Cove molding	1½ in. x 1½ in. x 42 in.	walnut
2	Cove moldings	1½ in. x 1½ in. x 21½ in.	walnut
1	Listel	¼ in. x 1⅜ in. x 39½ in.	walnut
2	Listels	¼ in. x 1⅜ in. x 20¼ in.	walnut
1	Torus-and-ogee molding	9/16 in. x ¾ in. x 40⅝ in.	walnut
2	Torus-and-ogee moldings	9/16 in. x ¾ in. x 20 13/16 in.	walnut

CUT LIST FOR TALL CHEST

Drawers

3	Third-width drawer fronts	$^{13}/_{16}$ in. x $4^{3}/_{16}$ in. x $12^{7}/_{16}$ in.	walnut
6	Third-width drawer sides	$^{1}/_{2}$ in. x $3^{15}/_{16}$ in. x $19^{7}/_{16}$ in.	poplar
3	Third-width drawer backs	$^{1}/_{2}$ in. x $3^{3}/_{16}$ in. x $11^{3}/_{16}$ in.	poplar
3	Third-width drawer bottoms	$^{3}/_{8}$ in. x $19^{1}/_{4}$ in. x $11^{5}/_{16}$ in.	poplar
2	Half-width drawer fronts	$^{13}/_{16}$ in. x $6^{1}/_{16}$ in. x $18^{13}/_{16}$ in.	walnut
4	Half-width drawer sides	$^{1}/_{2}$ in. x $5^{13}/_{16}$ in. x $19^{7}/_{16}$ in.	poplar
2	Half-width drawer backs	$^{1}/_{2}$ in. x $5^{1}/_{16}$ in. x $17^{5}/_{16}$ in.	poplar
2	Half-width drawer bottoms	$^{3}/_{8}$ in. x $19^{1}/_{4}$ in. x $17^{11}/_{16}$ in.	poplar
1	Drawer front	$^{13}/_{16}$ in. x $7^{1}/_{4}$ in. x $37^{7}/_{8}$ in.	walnut
2	Drawer sides	$^{5}/_{8}$ in. x 7 in. x $19^{7}/_{16}$ in.	poplar
1	Drawer back	$^{5}/_{8}$ in. x $6^{1}/_{4}$ in. x $36^{7}/_{16}$ in.	poplar
1	Drawer front	$^{13}/_{16}$ in. x $9^{1}/_{4}$ in. x $37^{7}/_{8}$ in.	walnut
2	Drawer sides	$^{5}/_{8}$ in. x 9 in. x $19^{7}/_{16}$ in.	poplar
1	Drawer back	$^{5}/_{8}$ in. x $8^{1}/_{4}$ in. x $36^{7}/_{16}$ in.	poplar
1	Drawer front	$^{13}/_{16}$ in. x $11^{1}/_{4}$ in. x $37^{7}/_{8}$ in.	walnut
2	Drawer sides	$^{5}/_{8}$ in. x 11 in. x $19^{7}/_{16}$ in.	poplar
1	Drawer back	$^{5}/_{8}$ in. x $10^{1}/_{4}$ in. x $36^{7}/_{16}$ in.	poplar
1	Drawer front	$^{13}/_{16}$ in. x $13^{1}/_{4}$ in. x $37^{7}/_{8}$ in.	walnut
2	Drawer sides	$^{5}/_{8}$ in. x 13 in. x $19^{7}/_{16}$ in.	poplar
1	Drawer back	$^{5}/_{8}$ in. x $12^{1}/_{4}$ in. x $36^{7}/_{16}$ in.	poplar
4	Muntins	$^{3}/_{4}$ in. x $1^{1}/_{4}$ in. x $19^{1}/_{4}$ in.	poplar
8	Drawer bottoms	$^{3}/_{8}$ in. x $19^{1}/_{4}$ in. x $17^{15}/_{16}$ in.	poplar

Hardware

5	Bail-type drawer pulls	$3^{1}/_{2}$-in. boring
8	Bail-type drawer pulls	4-in. boring

CONSTRUCTING THE TALL CHEST breaks down into four distinct phases. The first, of course, is constructing the case, and it is the phase with the most different operations. When the case is completed, you make the bracket feet and then add all the trim. The last phase is the drawer making. The sole complication here is that there are nine drawers in six different sizes.

MAKING THE CASE

Cutting the parts to rough size

The case is built using two different woods: walnut and poplar. Following tradition, the poplar is used wherever it won't be seen in the completed chest. So the top frame and the bottom panel, the drawer runners and guides, the rear drawer rails, and the back boards are all poplar. The remaining parts are walnut. With the exceptions of the side panels and the back boards, all the stock for the casework is the same ¾-in. thickness. To ensure that the stock is of uniform thickness, dress all of it at the same time.

Tip: Make the first cope cut on a strip of scrap. Tuck this coped scrap into the sticked edge of a rail before coping it, and the scrap will back up the workpiece, preventing the bit from blowing out splinters in the fragile profile.

1. Crosscut all the parts to their rough length. Put them in stickered stacks for a few days to acclimate to your shop.
2. Joint one face of each piece to produce a flat reference surface for thicknessing. Then joint one edge.
3. Thickness plane the stock.
4. Rip the stock to the desired width.
5. Label the parts on their ends. Store in stickered stacks until needed.

Photo A: A matched pair of bits makes quick work of stile-and-rail joinery. The sticking (or stile) cutter simultaneously slots and molds every rail and stile. The cope (or rail) cutter forms a tongue to fit the slot and at the same time cuts the negative of the molded profile.

Gluing up the panels

Rather than have a forgotten glue-up stall progress, I like to get any panels that will be required made up first thing. Unless you have unusually wide boards, you'll have to glue panels for the top and the bottom. And it is pretty likely you'll need to create wide boards for the side panels, for at least the bottom drawer, and perhaps for one or two other drawers.

1. Assess the widths of your boards and list the parts, based on the cut list, that you need to create by edge-gluing narrow boards.
2. Gather the boards for each panel.
3. Joint the mating edges.
4. Glue and clamp each wide panel. Make sure each is flat and true as you clamp it. Clean up squeeze-out immediately, in whatever method is appropriate for the glue you use. This will save scraping and sanding later.
5. When the glue has cured, scrape and sand the panels as necessary. Don't size them until you need them and can cut them to fit.

Making the side frames

The side frames are joined with contemporary cope-and-stick joinery. Cut the joints with a matched pair of router bits run in the table-mounted router (see **photo A**). Whether you do the cope cut first or the sticking cut is one of those pins-first or tails-first arguments. I do the sticking cut first so I get the profile cut to the depth that looks best. Achieving a properly fitted joint after this start isn't difficult.

1. Cut rails and stiles. The stiles should be 1 in. to 2 in. longer than specified by the cut list. The rail length specified allows enough material for a ⅜-in.-wide cope cut on each end.
2. Rout the sticking on one edge of each stile, on one edge of the top and bottom rails, and on both edges of each middle rail. Use the table-mounted router for this operation.
3. Cope the ends of the rails.

Raising the side panels

Raising a panel means a heavy cut for a router, even a high-horsepower one, so the cut should

SIDE ASSEMBLY LAYOUT

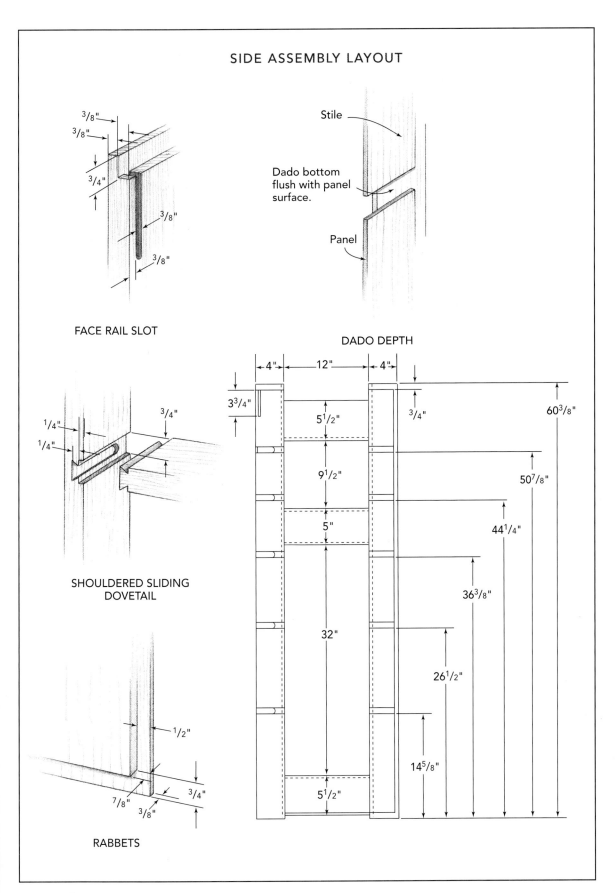

Stile

Dado bottom flush with panel surface.

Panel

FACE RAIL SLOT

SHOULDERED SLIDING DOVETAIL

RABBETS

DADO DEPTH

Photo B: When you rout the panels, take modest bites, so the router doesn't bog down. Clamp featherboards to the router-table fence to provide downforce on both infeed and outfeed sides of the bit.

Tip: Reduce the number of passes necessary to raise a panel by first beveling the edges of the workpieces on the table saw.

be completed in three or four passes. You can begin with the bit low and raise it before each pass, or you can set the final bit height and moderate the cut by altering the fence position before each pass (see **photo B**).

1. Cut the side panels to final size.
2. Set up the router table. Chuck the panel-raising bit in the router and raise the bit so the pilot bearing is slightly proud of the tabletop. Position the fence flush with the bearing. By making a series of test cuts on a spare piece of the working stock, raise the bit to the final cut depth. This is the point at which the panel's tongue fits perfectly in the frame's panel groove.
3. Raise the panels in increments by adjusting the fence position, rather than by altering the bit height. To begin, move the fence well forward of the pilot. Cut all the panels. With each one, make the first cut across the end grain; that way you'll end with a long-grain cut. Move the fence back and cut all the panels. For the final cut on all panels, the fence

should once again be flush with the pilot bearing.
4. Apply your finish to the panels now. You need to do this before assembly to avoid having a sliver of unfinished wood exposed when the panels shrink seasonally. Apply the finish to both sides and all edges equally. I used an oil finish.

Assembling the sides

Assemble both sides without glue first. This is the opportunity to check how all the joints fit as well as to rehearse the assembly sequence, get clamps set, and ready the cauls that will protect the work from the clamp jaws. Lightly mark the rail locations on the stiles, so you can position the rails accurately after the glue is applied. When everything is checked, open the glue.

1. Glue the middle rail into one stile. Use a brush to apply glue to the stub tenons on the rail and press it into the panel groove. Avoid sliding the rail back and forth in the groove to

Photo C: Assembling a side is straightforward. Apply glue with a brush to confine it to the surfaces that require it. Join the middle rail to a stile, add the panels—but don't glue them—and then add the top and bottom rails and the second stile.

get it aligned, since this will smear glue where you don't want it.

2. Fit the panels into place. Remember that the panels are not glued. They need to be able to expand and contract.

3. Glue the top and bottom rails into the stile (see **photo C**). Again, apply glue carefully with a brush and avoid transferring any of it to the panel edges.

4. Set the second stile in place. Check to make sure the panel is square and flat.

5. Apply clamps; recheck for square and flat.

Cutting the sides joinery

Web frames tie the sides together to form the case. At the same time, they support the drawers. The lower frames, which hold the full-width drawers, consist of a front rail and two runners; no back rail is needed because the runners are fastened to the sides at the back. The two frames that support the half- and third-width drawers do have back rails, because their intermediate runners need support at the back (see "Web Frame Construction" on p. 198).

The rails and runners are joined with groove-and-stub-tenon joinery. The frames formed by the rails and runners are housed in shallow dadoes cut into the stiles of the side frames. In addition to the dadoes, the front rails lock into the sides with dovetails. At the back, as mentioned, the runners for the full-width drawers are fastened to the sides.

The trick is centering dovetail slots in the dadoes. I resolved this challenge by using two identical routers, one with a straight bit for the dadoes, one with a dovetail bit for the dovetail slots. I set up a guide and cut the dado and then the slot (see **photo D** on p. 199).

In addition to the cuts for the web frames, the sides must be rabbeted for the top frame, bottom, and back and grooves must be cut for the face rail.

1. Cut the web frame dadoes first. The dado bottoms are flush with the backs of the raised panels. However you choose to cut these dadoes, set your cutter to dado the stiles without cutting the panels. Cut all dadoes in both

Web Frame Construction

NOTE THAT THE DRAWER GUIDE fits between the stiles of the side assembly and are glued to the runner only (not to the panel). The runner tenons are glued into the slot in the front rail and fastened with brads.

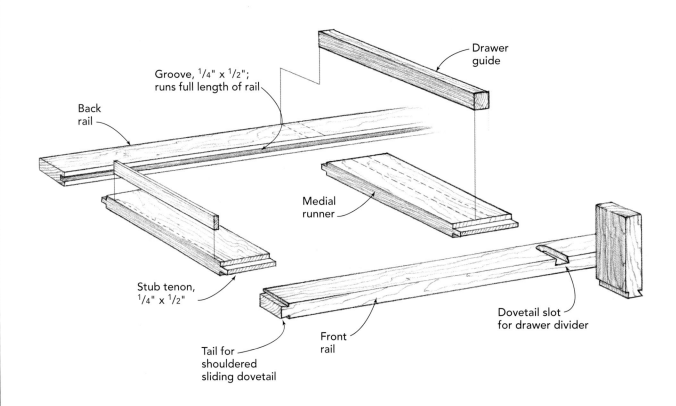

Drawer guide

Groove, $^{1}/_{4}$" x $^{1}/_{2}$"; runs full length of rail

Back rail

Medial runner

Stub tenon, $^{1}/_{4}$" x $^{1}/_{2}$"

Tail for shouldered sliding dovetail

Front rail

Dovetail slot for drawer divider

SINGLE-DRAWER WEB FRAME

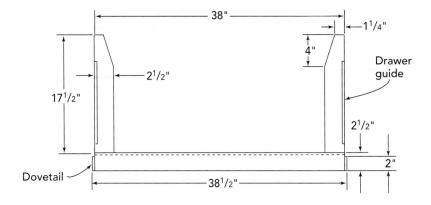

38"

$1^{1}/_{4}$"

4"

$2^{1}/_{2}$"

$17^{1}/_{2}$"

Drawer guide

$2^{1}/_{2}$"

2"

Dovetail

$38^{1}/_{2}$"

TOP WEB FRAME

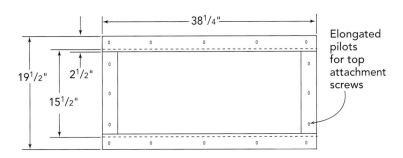

38¹/₄"

19¹/₂"   2¹/₂"

15¹/₂"

Elongated
pilots
for top
attachment
screws

DOUBLE-DRAWER WEB FRAME

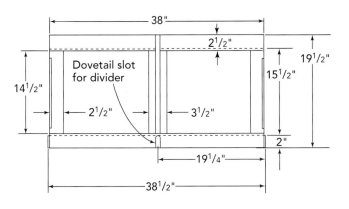

38"

2¹/₂"

Dovetail slot
for divider

19¹/₂"

15¹/₂"

14¹/₂"

2¹/₂"   3¹/₂"

2"

19¹/₄"

38¹/₂"

TRIPLE-DRAWER WEB FRAME

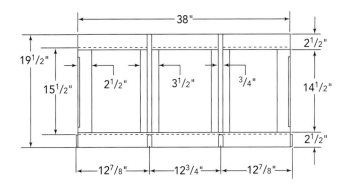

38"

19¹/₂"   2¹/₂"

15¹/₂"   2¹/₂"   3¹/₂"   ³/₄"   14¹/₂"

2¹/₂"

12⁷/₈"   12³/₄"   12⁷/₈"

front and back stiles of each side assembly as
shown in "Side Assembly Layout" on p. 195.
2. Cut the ¹/₄-in.-deep by 2-in.-long dovetail
slots. Cut from the front edge of the side in.
The slots must be centered inside the dadoes.
3. Cut rabbets for the top and bottom. The
rabbets are ³/₄ in. wide, the same as the thickness of the top and bottom, and ³/₈ in. deep.
Because the stiles extend several inches
beyond the top side rail, the rabbet for the top
cuts only through them. At the bottom, the
rabbet does catch the edge of the bottom rail.
4. Cut the rabbet for the back boards. This
rabbet is ¹/₂ in. wide (matching the thickness of
the back boards) and ³/₈ in. deep.
5. Cut the groove for the face rail. The groove
is ³/₈ in. wide and ³/₈ in. deep, and 3³/₄ in. long.

Making the web frames

The main purpose of the web frames is to support the drawers. At the same time, they tie
the two side assemblies together, forming the

**Photo D: Having two identical routers, one with a straight bit, one with
a dovetail bit, allows you to cut the dado and the dovetail slot using
the same guide setup.**

Photo E: Cutting the shouldered dovetail is a simple router-table operation. Make test cuts on scrap to dial in the proper bit height and depth.

Photo F: A large-diameter mortising bit in a router table cuts the stub tenons quickly. The fence controls the length of the tenon. Guide the work along the fence with a square-edged back-up block.

case. In this project, there's also a top frame, which is the structural top of the case. The chest top—a panel—is attached to the chest by way of this frame.

The tall chest has, including the top frame, five different web frame configurations, four of which you make now (see "Web Frame Construction" on p. 198).

The fifth frame is for the bottom drawer; it consists of a front rail and two runners that are nailed to the case bottom. Making and installing these parts is one of the last things you do in the whole project.

1. Cut the web frame parts. These include the front and back rails, the runners, and the guides. Note that there is a lot of variety in the frames. You will cut rails of slightly different lengths and runners of different widths and lengths. Some parts are cut from the primary wood, though most come from the secondary wood. As you cut the parts, label them clearly.

2. Cut tails on the ends of the front drawer rails (except the very top and very bottom ones). I cut these on the router table, with the same bit I used to cut the slots (see **photo E**). The tails need to be fitted carefully to the slots routed in the sides. Make test cuts on scraps of the working stock to home in on the perfect bit height and depth. The tails will need to be trimmed, but you can do this after grooving the rails.

3. Cut grooves in the "inside" edges of all the rails. The grooves are ¼ in. wide and ½ in. deep. Center them on the edge. You can do this on the table saw or router table.

4. Trim the dovetails on the front rails to the length of the dovetail slots. Check the length against the slot without driving the tail completely into the slot.

5. Cut the stub tenons on the runners. Remember that only the short runners are tenoned on both ends. Cut the tenons to fit the grooves in the rails. I cut the tenons on the router table with a large-diameter mortising bit (see **photo F**).

6. Chamfer a back corner off of each long runner.

WOOD MOVEMENT AND FRAME-AND-PANEL CONSTRUCTION

Furniture makers always have to be mindful of wood movement. A 20-in.-wide red oak panel for a chest side, for example, might move as much as ⅜ in. between summer and winter. Even a relatively stable wood like walnut will move as much as ⁵⁄₁₆ in. So how do you construct a chest of drawers so this movement won't tear the piece apart?

Frame-and-panel construction is one good approach. The idea is to build a frame around a wide panel. The frame, because it is made up of narrow members, doesn't change much. Though the panel does move quite a bit, it's set into the frame in such a way that it can expand and contract without bursting the frame.

In the typical simple frame, two rails (the horizontals) are trapped between two stiles (the verticals). Because normal wood doesn't move along its grain, the frame won't elongate and shrink, it maintains its length. Where the change in dimension comes is across the frame; the stiles will expand and contract across their grain. But the stiles in the tall chest, for example, are 4 in. wide and, with a wood of average stability like walnut, they'll move about ³⁄₆₄ in. across each stile. Maxed out, that 18-in.-wide frame-and-panel assembly, will expand only ³⁄₃₂ in. A lot more manageable than ⁵⁄₁₆ in.

The panel, being roughly 13 in. wide, will expand and contract. But set in grooves in the edges of the frame, it can expand and contract without changing or damaging the frame.

---

7. Assemble the web frames. Glue the runner tenons into the grooves in the rails. Make sure the assemblies are flat and square. Pin the tenons in place with a brad or two.

8. Cut dovetail slots for the vertical drawer dividers in the two full frames. The slots are the same size as those cut for the drawer rails. The uppermost web frame (not the top frame) has two slots in the upper surface of its front rail and one in the underside. The other frame has a single slot in its front rail's upper surface as shown in "Web Frame Construction."

9. Fasten the guides to the upper two web frames. These ¾-in.-square strips of wood extend from the seam between the runner and the front rail to the back of the frame. Glue the guides to the intermediate runners.

Beginning the case assembly

You need to begin assembling the case so you can accurately determine the lengths of the top frame, the bottom, and the face rail. What you do is join the sides with two or three of the web frames and then take your measurements. Cut the remaining case parts, make the remaining joinery cuts; then complete the assembly.

1. Set up the sides on a solid assembly table.

2. Apply glue to the rear dadoes for a couple of the web frames. Apply a sparing amount also to the dovetails on the web frames themselves.

3. Slide the first frame into place. As the dovetails enter their slots, gently drive the frame down until the front edge of the rail is flush with the edges of the sides (see **photo G** on p. 202).

4. Slide the second frame in place.

5. Apply bar or pipe clamps across the back. Tighten the clamp enough to hold the web frame, but not to distort the case. The long

Tip: Trying to balance a side (or worse, two sides) on edge while you assemble a case can make you nuts. Wood handscrews clamped on a side— one on each end— can serve as feet to keep it from tumbling onto its face during initial assembly work.

Photo G: Prop up the sides on an expansive assembly table to install the first of the web frames. Installed, the first frame ties the sides together and simplifies the addition of the rest of the frames.

4. Lay out and cut elongated pilot holes in the bottom panel. These holes are for mounting the bracket foot blocking on the sides of the chest (see "Bracket Foot Construction" on p. 205). For each piece of blocking, there is a round pilot drilled for the fixed point. The elongated pilots allow the bottom panel to move without breaking the glue joints between the blocking and the posts. These pilots should be counterbored on the inner face of the bottom, so the bottom drawer runners can be placed over the screw heads.

5. Rip and crosscut the face rail.

6. Form a bare-faced tenon on each end, by cutting a ⅜-in.-wide by ⅜-in.-deep rabbet across one face and one edge (a bare-faced tenon has one cheek—in this case, the back one—flush with the surface of the workpiece). I cut this tenon on the router table. Check the fit of the face rail in the case.

7. Lay out and cut the mortises for the kickers. The kickers are roughly centered above each opening in the top row of drawers; the exact lateral position isn't critical. But the bottom edge of the kicker should be flush with the bottom edge of the face rail. The mortises should be ¼ in. by 1 in. and ¼ in. deep. The mortises can be routed or simply excavated with a chisel.

Making the kickers

A kicker is a case component that keeps a drawer level as it is opened. In this chest, the runners double as kickers for the drawer below them. Because there is no web frame immediately above the top row of drawers, kickers are needed for them.

1. Measure the case for the kickers. They extend from the face rail to the back board. A stub tenon joins each kicker to the face rail, while nails driven through the back board secure it at the back. Measure from the inner face of the face rail to the shoulder of the back rabbet. Add ¼ in. (the length of the tenon) to this measurement.

2. Cut the kickers from the secondary stock. Plane, if necessary, and rip stock to ¾ in. by

runners must be screwed or nailed to the case sides, since they are not supported by a back rail.

Making the top frame, bottom, and face rails

1. Measure from side to side for the top frame, the bottom panel, and the face rail.

2. Crosscut the top frame and and bottom panel to fit.

3. Lay out and cut elongated pilot holes in the top frame. The pilots are for screws that attach the top panel to the chest as shown in "Web Frame Construction" on p. 198. In the front rail, the pilots are simple drilled holes, because the top is fixed at that point. In the runner and the back rail, the pilots are elongated so the top can move. These pilots should be ⅛ in. wide and ⅜ in. long.

1½ in. Crosscut three pieces of this stock to the length just determined.

3. Cut the tenon on one end of each kicker. The tenon is ¼ in. by 1 in. by ¼ in.

Making the vertical drawer dividers

The vertical dividers separate the individual drawers in the two top rows.

1. Cut the dividers, which were roughed out from the primary stock. The length of the divider that fits between the two web frames is critical, but the length of the two dividers for the top row of drawers is not.

2. Rout dovetails on both ends of the short divider and on one end of the longer ones. The dovetails must fit the slots routed in the front rails.

3. Trim the dovetails to match the length of the slots.

4. Notch the front corners of the top two dividers to accommodate the face rail.

Completing the case assembly

Now that you've made the remaining case parts, you can finish assembling the case.

1. Glue the top two vertical drawer dividers into their slots in the top web frame.

2. Apply glue to the tenons on the face rail and slide it into place.

3. Drive a screw or nail through the top end of each divider into the back of the face rail (see **photo H**).

4. Glue and nail the top frame and bottom panel to the sides. Remember that these parts are flush with the shoulders of the back rabbet, so the back boards can overlap them. Drive the nails through the face of the panels into the shoulder of the rabbets in the sides. You can also drive three or four nails through the top into the edge of the face rail. I used cut nails, and I did drill pilot holes before hammering them in.

5. Glue the remaining web frames into the case. Nail or screw the runners into the dadoes.

Photo H: With a clamp holding the parts, screw the drawer divider to the face rail. One screw angled through the divider's top end is all that's needed.

Photo I: Glue the kickers into their mortises just before setting the top back board in place. Nail the board to the case and to the ends of the kickers.

6. Glue the second-row drawer divider into its slots.

7. Glue the guides to the side runners. These very thin strips fit between the stiles of the side assemblies, to prevent a drawer from getting slightly cocked. You must be careful to glue the guide strips to the runner and not to the panel.

Constructing the back

The back is made of random-width boards, jointed edge to edge with shiplap joints. You can close in the case back now, or you can install the top back board (so the kickers can be mounted) and leave the back open while

you construct and mount the feet, and make and fit the drawers. I did the latter.

1. Prepare the stock if you haven't already done so. Use your secondary wood, planing it to a ½-in. thickness. The widths of the boards are random; just rip the stock to maximize the yield. Crosscut the boards to fit the space from rabbet to rabbet across the back.

2. Cut the shiplaps on all boards. A shiplap joint is formed by nesting two rabbeted edges. For this chest, cut ½-in.-wide rabbets in the long edges of all but two of the boards. The depth of the rabbet should be one-half the board's thickness. Cut the rabbets into opposite faces; that is, when you look at the board face on, you should see only one rabbet. The other rabbet is cut in the other side. The top and bottom board each have one square edge and one rabbeted edge.

3. Glue the kickers into the mortises in the face rail (see **photo I**).

4. Nail the uppermost back board to the case. Use cut nails. The board overlaps the back edge of the top frame and is flush with its top surface. The rabbet faces in.

5. Align each kicker; then drive a couple of nails through the back board into its end. Secure all three kickers.

MAKING THE BRACKET FEET

What you see in a bracket foot is an attractive facing that does little to support the chest. Hidden behind the facing is a post. which is really holding the chest up off the floor.

Despite an appearance of strength, the traditional construction is problematic because the critical gluing surfaces are cross-grain. I used a design that eliminates the cross-grain construction.

Making the molded facing

The most challenging part of the construction is making the facings. You cut a large-scale ogee profile on a heavy strip of the primary wood.

1. Joint and plane enough 1¾-in.-thick stock to make the facings. To begin, you need a board wider than the foot is high and about 10 times longer than its length. After the board is shaped, you'll cut rip it to final width (6 in.) and crosscut it into six pieces. The extra length is cheap insurance. It is easier, I think, to work with a couple of short pieces than one long one, so I used a couple of 48-in. pieces.

2. Make a cardboard pattern of the profile of the molded facing (see "Molding Bracket Feet" on p. 206). For now, lay out the boundaries of the desired cove on a setup blank, as indicated in the drawing. Later, after the cove is cut, you can trace the pattern for the shoulder onto the ends of the workpieces.

3. Set up the table saw for the cove cut, as detailed in "Cove Cutting on the Table Saw" on p. 210.

4. Plow the cove. Lower the blade and make a pass, guiding the stock against the fence. Cut

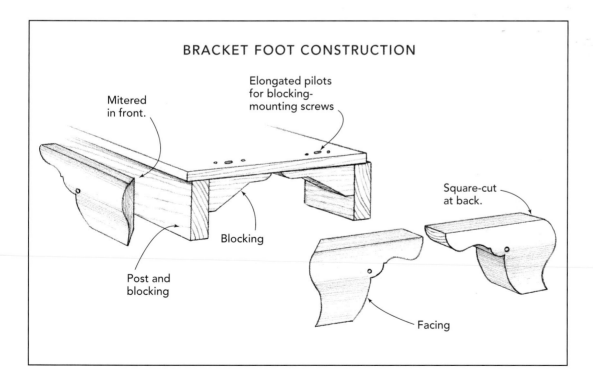

BRACKET FOOT CONSTRUCTION

Mitered in front.

Elongated pilots for blocking-mounting screws

Square-cut at back.

Blocking

Post and blocking

Facing

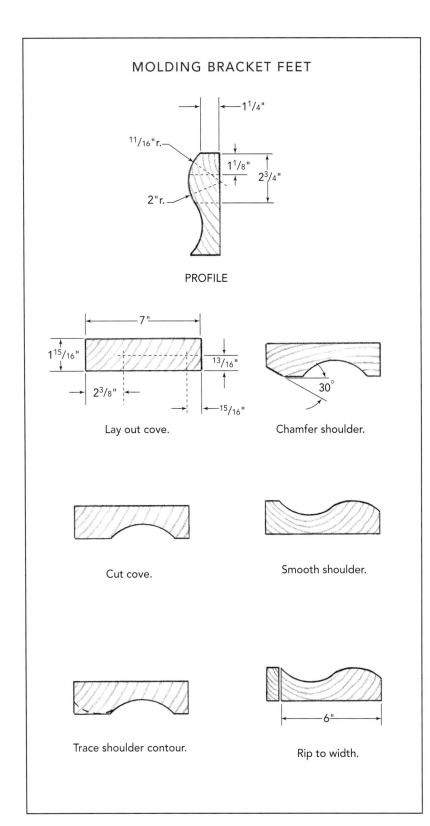

MOLDING BRACKET FEET

1¼"

11/16" r.

1⅛"

2¾"

2" r.

PROFILE

7"

1¹⁵/₁₆"

13/16"

2³/₈"

15/16"

Lay out cove.

30°

Chamfer shoulder.

Cut cove.

Smooth shoulder.

Trace shoulder contour.

6"

Rip to width.

Photo J: Chamfer off the bulk of the waste on the table saw; then shape the shoulder with a spokeshave or plane. Smooth the surface with coarse sandpaper or a curved scraper.

each workpiece. Raise the blade about ⅛ in. and make another pass on each workpiece. Make additional shallow passes until you've completed the cove.

5. Bevel the shoulder. The radius of this curve is far too big to be made with a router bit. Instead, make one or more bevel cuts on the table saw to trim away waste and bring the contour close to the outline on the stock's end.

6. Shape the shoulder contour with hand tools. Use a plane, a spokeshave, rasps and files, and scrapers to shape the shoulder to match the laid-out contour (see **photo J**).

7. Smooth the profile. Clean up the molding with a scraper and sandpaper.

Cutting the facing pieces

1. Cut six facings from the strips of ogee molding.

2. Miter one end of the four pieces that form the two front feet. The side facings for the back feet are square cut.

3. Draw full-size patterns on posterboard, one pattern for each cutout (see "Bracket Foot Layout"). Bear in mind that there are three different lengths used. The pieces on the front are the longest. The piece on the side at the back is the shortest.

4. Because their faces are shaped, you need to rest the blanks on their backs to cut them on the bandsaw. Trace the appropriate pattern onto each facing blank.

5. Cut out the foot facings on the bandsaw.

Making the feet and blocking

The chest rests on four posts hidden by the facings. These posts are a common failure point in old chests, because the post that really supported the chest had its grain oriented vertically, making it cross-grain to the blocking glued to either side to support it as well as to the facing. This tall chest has the post integrated into part of the blocking, and the grain of this piece runs horizontally, rather than vertically. When glued to the remaining blocking and the facings, all of which have their grain oriented horizontally, the bracket foot is strong and rigid.

1. Cut the posts and blocking (see "Bracket Foot Construction" on p. 205). Mill 10/4 stock to a 2 in. thickness. Rip and crosscut the blanks, layout the contours, and bandsaw the parts.

2. Glue the post-and-blocking pieces to the bottom of the case. The "post" ends of the pieces should align on the seam between the bottom and the sides. In the front, the faces should be inset ⅜ in. In the back, the faces are flush with the case back. You can drive nails or screws through the case bottom into the post-and-blocking pieces.

3. The blocking is mounted next. To avoid cross-grain problems, don't glue these pieces

to the case bottom. Glue and screw them to the post-and-blocking pieces (see **photo K** on p. 208). Run pan-head screws through the pilot holes in the bottom into the blocking.

4. Mount the molded faces. Glue them to the blocking.

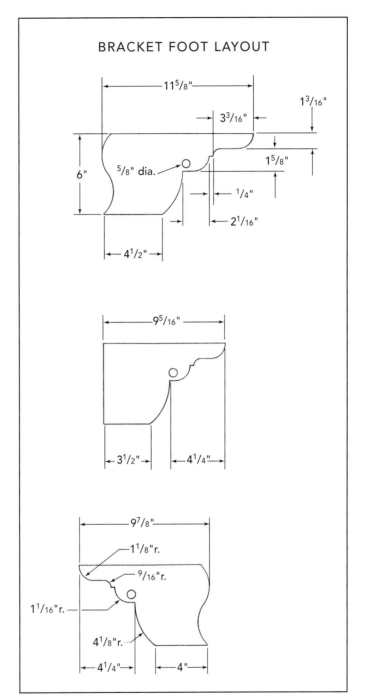

BRACKET FOOT LAYOUT

Photo K: The one-piece posts and blockings—the main supports for the chest—are glued to the bottom of the piece. To brace them, blocking is screwed to the posts and to the chest bottom.

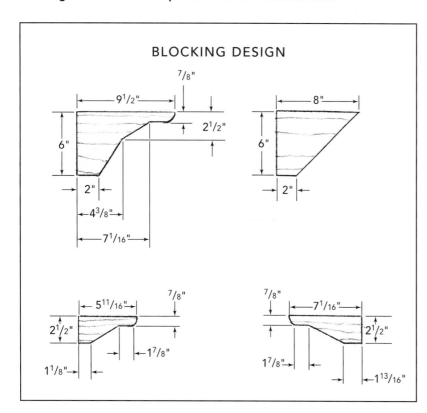

BLOCKING DESIGN

TRIMMING OUT THE CASE

The tall chest is embellished with two moldings, a simple base molding and a built-up cornice molding. I cut both moldings on the table saw and with a router and then glued them to the chest. The profiles are shown in "Molding Profiles."

Making and installing the base molding

The base molding is a simple profile, meaning it has a single geometric shape: an ovolo. Cut it in a pass or two on the router table.

Before you can install the molding, you have to install the bottom drawer rail and runners. These parts were set aside, so you'd have unrestricted access to the case bottom for driving mounting screws into the bracket feet.

1. Glue the drawer rail to the case bottom.
2. Nail or screw the runners to the case bottom. If you screw the runners, drill a fixed pilot for a screw close to the rail; then make elongated pilots for screws at the middle and back end of the runner. This will allow the case to move seasonally.
3. Cut a piece of the primary stock. Make it ¾ in. by 2½ in. by 42 in.
4. Set up the router table to cut the profile. I used a window-trim bit (Jesada #604-600; see "Sources" on p. 216). It's a roundover with an elliptical curve that is ¾ in. high and ⅝ in. wide.
5. Mold the two long edges of the blank.
6. Rip the blank into two molding strips, each measuring ¾ in. by 1 in. One strip will be the front molding, the other is cut in half for the sides.
7. Miter the ends of the front molding to fit the chest. Glue this strip to the case and the top ledge of the bracket feet.
8. Miter the appropriate end of each side molding strip. Glue them to the side stiles and the top ledges of the bracket feet. The molding will cover (finally) the small gap between the stiles below the bottom side rail.

MOLDING PROFILES

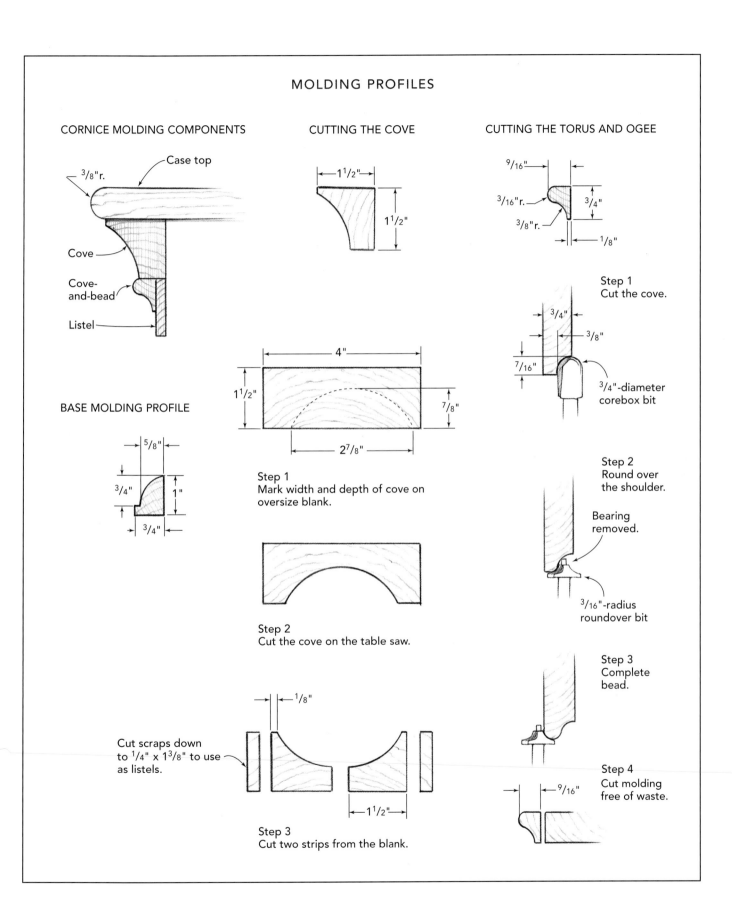

CORNICE MOLDING COMPONENTS

Case top

3/8"r.

Cove

Cove-and-bead

Listel

CUTTING THE COVE

1 1/2"

1 1/2"

4"

1 1/2"

7/8"

2 7/8"

Step 1
Mark width and depth of cove on oversize blank.

Step 2
Cut the cove on the table saw.

1/8"

Cut scraps down to 1/4" x 1 3/8" to use as listels.

1 1/2"

Step 3
Cut two strips from the blank.

BASE MOLDING PROFILE

5/8"

3/4"

1"

3/4"

CUTTING THE TORUS AND OGEE

9/16"

3/16"r.

3/8"r.

3/4"

1/8"

Step 1
Cut the cove.

3/4"

3/8"

7/16"

3/4"-diameter corebox bit

Step 2
Round over the shoulder.

Bearing removed.

3/16"-radius roundover bit

Step 3
Complete bead.

Step 4
Cut molding free of waste.

9/16"

COVE CUTTING ON THE TABLE SAW

1. Set the blade height. Sketch the approximate cove profile on a test piece of the working stock. Set it beside the blade and adjust the blade height (see photo at near right).

2. Find the fence angle. Adjust your shopmade parallel rule (see drawing below) to the width of the desired cove. Straddle the blade with the rule. Find where the blade touches both straightedges simultaneously. When you've found the angle, mark along the straightedges on the saw table (see photo at top far right).

3. Offset the layout lines to accommodate a workpiece that's wider than the cove itself.

4. Clamp parallel fences to the saw table. Cut the cove. Raise the blade only ⅛ in. above the saw table. You need to provide downward pressure on the work at the same time you are advancing it. After the first pass, raise the blade another ⅛ in. for a second pass, and again for a third and fourth pass, as necessary (see photo at bottom far right).

5. Sand the cove. It definitely will need smoothing.

PARALLEL RULE PLAN

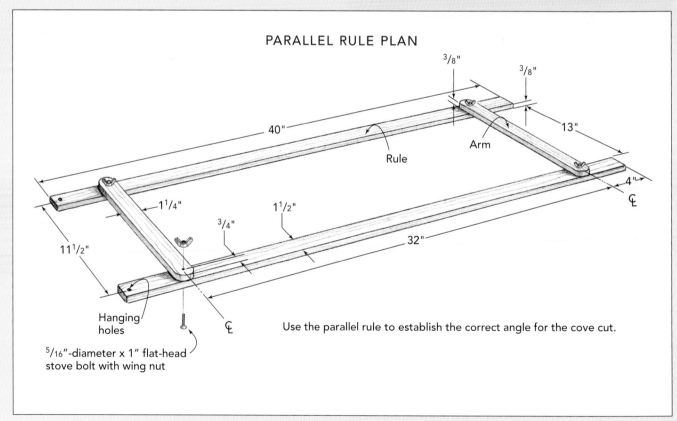

Use the parallel rule to establish the correct angle for the cove cut.

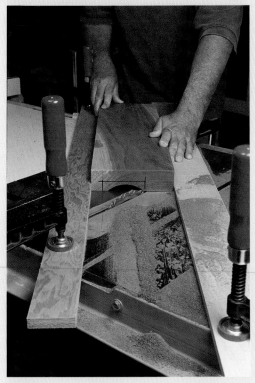

Photo L: For the torus-and-ogee molding, you want to blend the roundover profile into the cove seamlessly. Remove the pilot bearing from the roundover bit and control the placement of the profile on the blank with the router-table fence.

9. Trim the ends of the base molding flush with the back of the chest.

Making the cornice molding

The cornice is composed of several profiles and four separate pieces of wood. The top of the chest has a bullnose (or torus) machined on the front and sides. Tucked beneath the overhang of the top is a fairly large (1⅜-in.-radius) cove molding. Butted against the bottom of the cove, tight to the chest side is a thin strip called a listel. A molding with a torus and reverse ogee profile fits over the listel, up against the bottom of the cove molding. Cut the strip moldings now.

1. Cut the blanks. You need two blanks for the cove molding and one wide stick for the torus-and-ogee molding. The listel is ready to use as it leaves the saw, but for a pass or two with a plane or sandpaper.
2. Cut the cove molding on the table saw, as you did the ogee moldings for the bracket feet.
3. Rip the cove block to final dimensions.
4. Rout the torus-and-ogee molding on the router table (see **photo L**). Follow the sequence shown in the drawing on p. 209.

Tip: It is always best to mold substantial sticks, rather than risk having a fragile strip splinter or virtually explode as you rout it. Rout the profile on one or both edges of a 3-in.- or 4-in.-wide board; then rip the molded edge from it.

Tip: With solid-wood case sides, the molding can't be glued across the grain. The movement of this chest's post-and-panel side assembly is nil. Thus gluing the molding to the stiles is unlikely to be a problem.

Making the top

The top is a key component of the cornice. It is probably one of the first parts you made. Now it is time to complete it and attach it to the chest.

1. Measure the chest and the cove molding and confirm the dimensions of the top.
2. Rip and crosscut the top to size.
3. Machine the bullnose on the front and sides of the panel.

Mounting the top and cornice

1. Attach the top with screws driven up through the top frame. You already made the pilot holes in the frame. The top is fixed at the front and can expand to the back seasonally.
2. Miter the ends of a length of the cove molding so it fits across the front of the case. Glue this piece to the face rail and the underside of the top.
3. Fit lengths of the listel and the torus-and-ogee molding to the case front. Miter the ends. Glue the listel to the case, and the molding to the listel and the bottom edge of the cove (see **photo M**).

4. Attach the moldings to the chest side. Miter an end of the cove, the listel, and the torus and ogee. Glue them to the chest side but not the top. Then trim the moldings flush with the back of the chest.

BUILDING THE DRAWERS

Although there are quite a few different sizes of drawers in this chest, all are made in basically the same way. Routed half-blind dovetails join the sides and fronts, the back fits in a dado in each side. The bottoms are solid wood, rabbeted to fit the usual bottom groove in the sides and front (see "Drawer Construction").

The differences that do come into play are that the full-width drawers are made with ⅝-in.-thick sides and backs. And a center muntin strengthens the bottom construction. Because they are much smaller, the half- and third-width drawers are built with ½-in.-thick sides and backs.

Good at production work? Then mill stock and cut all the parts for all the drawers at one time. Less confident? Do the smaller drawers

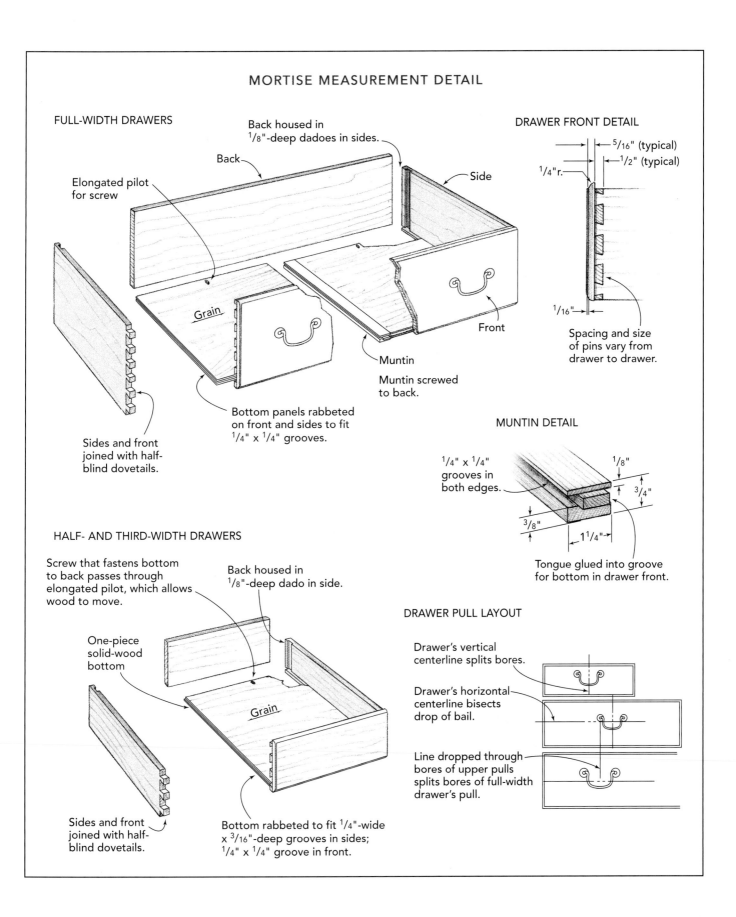

MORTISE MEASUREMENT DETAIL

FULL-WIDTH DRAWERS

Back housed in ¹/₈"-deep dadoes in sides.

Back

Elongated pilot for screw

Side

Grain

Front

Muntin

Muntin screwed to back.

Bottom panels rabbeted on front and sides to fit ¹/₄" x ¹/₄" grooves.

Sides and front joined with half-blind dovetails.

DRAWER FRONT DETAIL

⁵/₁₆" (typical)

¹/₂" (typical)

¹/₄" r.

¹/₁₆"

Spacing and size of pins vary from drawer to drawer.

MUNTIN DETAIL

¹/₄" x ¹/₄" grooves in both edges.

¹/₈"

³/₄"

³/₈"

1¹/₄"

Tongue glued into groove for bottom in drawer front.

HALF- AND THIRD-WIDTH DRAWERS

Screw that fastens bottom to back passes through elongated pilot, which allows wood to move.

Back housed in ¹/₈"-deep dado in side.

One-piece solid-wood bottom

Grain

Sides and front joined with half-blind dovetails.

Bottom rabbeted to fit ¹/₄"-wide x ³/₁₆"-deep grooves in sides; ¹/₄" x ¹/₄" groove in front.

DRAWER PULL LAYOUT

Drawer's vertical centerline splits bores.

Drawer's horizontal centerline bisects drop of bail.

Line dropped through bores of upper pulls splits bores of full-width drawer's pull.

first; when you complete them, then do the large ones. In either approach, the drawer construction follows the same routine.

Cutting the parts

1. If you haven't already done so, joint and thickness plane stock for the drawer fronts. Prepare stock for the sides, backs, muntins, and bottoms as well. For the two lowest drawers and for the drawer bottoms, you may need to glue up stock.

2. Rip and crosscut the drawer fronts. Before turning on the saw, compare the dimensions specified by the cut list with the dimensions of the drawer openings of your chest. Each drawer front should be ⁷⁄₁₆ in. longer than the opening and ³⁄₁₆ in. wider than the opening is high. Make sure the fronts you cut will fit your chest.

3. Rip and crosscut the sides and backs. As with the fronts, reconcile the dimensions specified on the cut list with the actual dimensions of your chest before cutting any of the parts. Bear in mind that the higher the drawer, the more space you need to allow for seasonal wood expansion. As you cut the parts, be sure to label them.

Cutting the grooves

A variety of joinery cuts are essential. The widths and depths of these cuts, and their locations are the same, regardless of the size of the drawer. That makes it possible to make each cut on the parts for every drawer with a single, common setup.

As usual, there are a number of ways you can make the cuts. I did them all on the router table.

1. Cut a groove for the bottom in the front; the sides; and, for some drawers, the muntin. The groove is ¼ in. wide by ¼ in. deep. Its bottom shoulder is ⅜ in. from the bottom edge of the sides and fronts (or in the case of the muntin, the bottom face).

2. Dado the sides for the back. The width of the dado depends, of course, on the thickness of the back. It is ⅛ in. deep and is ½ in. from the back end of the side. Cut it into the same

face as the bottom groove. On the router table, it's possible to make this dado a stopped cut, ending at the bottom groove. If you cut them through (from top edge to bottom edge), that's okay too.

3. Rabbet the drawer fronts. As I pointed out earlier, the fronts are rabbeted on three edges—the top and the ends. The small drawers and the shortest of the full-width drawers can be rabbeted the same on all three edges, that is ¼ in. wide and a ⁷⁄₁₆ in. deep. The taller full-width drawers should get a wider rabbet on the top edge so the drawer doesn't stick in protracted humid weather. Check the fit of the fronts in the actual drawer openings.

4. Rabbet the muntin. The front end of the muntin is rabbeted top and bottom, forming a tongue that fits into the bottom groove already cut in the drawer front.

5. Rabbet the drawer bottoms. The goal here is to create a tongue that fits the bottom grooves cut in the sides and fronts.

Cutting the half-blind dovetail

The drawer fronts and sides are joined with half-blind dovetails. Cut them however you prefer. I routed them using a Leigh jig, so they don't have the regular spacing common to routed half-blind dovetails. To rout half-blind dovetails in rabbeted drawer fronts, you must do the tails and the pins in separate operations, regardless of the jig you use.

Make test cuts to set up the dovetail jig, and get the fit right. With rabbeted drawer fronts, it is difficult to sand protruding pins flush.

Assembling the drawers

With the joinery cut, only one operation remains before assembly: that is beading the drawer-front edges. Do that, then assemble the drawers.

1. Round over the edges of the drawer fronts with a ¼-in.-radius roundover bit. Cut deep enough to leave a tiny shoulder. Rout all four edges of each drawer front.

Tip: *It is far easier to align a drawer front in a dovetail jig if all its edges are square. Save yourself some trouble and wait until after you've cut the dovetails to bead the fronts.*

Photo N: The muntin used to bolster the bottom of the full-width drawers is glued to the drawer front and back but not the to bottom panels. Use the first panel to position the muntin; then slide the second panel into place.

2. Run through a dry-assembly before actually gluing up each drawer. If the parts fit tightly enough to allow it, fit the drawer in the appropriate opening in the chest. You need to deal with fit problems and solve them before gluing up the drawers.

3. Drill or rout an elongated pilot for the lock screw that goes through the bottom into the drawer back at this time.

4. Assuming all goes well, glue up the drawer. After gluing the sides to the fronts, set the assembly upside down on a flat surface. Apply glue and slide the back into the dadoes and clamp it. Slide the bottom in place without glue and drive the locking screw. If the drawer has a muntin, it is glued into the drawer front and screwed to the back (see **photo N**).

Drilling the mounting holes

I used two sizes of pulls. Same pattern, but scaled to the proportions of the drawers. The full-width drawers have 4-in. borings, while the smaller drawers have 3½-in. borings.

On the smaller drawers, the pulls are centered left to right. On the large drawers, the outer boring is 3⅛ in. from the edge of the drawer front. On all the drawers, the horizon-

tal centerline splits the difference between the borings and the bottom of the bail.

1. Make a drilling guide for each size of pull that you'll need.

2. Lay out the pull locations on the drawer front.

3. Use the guide and a drill to bore holes for the mounting screws.

4. Check the fit of the pulls, but don't mount them until after the chest is completed and finished.

FINISHING UP

Two big jobs remain. You have to install the back and you have to apply a finish.

1. Mount the remaining back boards. Nail through the edges into the shoulder of the rabbets in the case sides. Rip the bottom board if necessary to make it fit.

2. Shellac the drawers (except, of course, for the exposed face of the fronts.

3. Apply a finish to the chest.

4. Mount the pulls.

SOURCES

JULIUS BLUM, INC.
7733 Old Plank Rd.
Stanley, NC 28164
(800) 438-6788
www.blum.com
Manufacturer of hardware for cabinets and furniture, including drawer runners

CERTAINLY WOOD
13000 Route 78
East Aurora, NY 14052-9515
(716) 655-0206
www.certainlywood.com
Sells veneers

FISHER FORGE
David W. Fisher
150 Fisher Ct.
Hamburg, PA 19526
(610) 562-5425
Maker of hand-forged hinges and other hardware

FREUD, INC.
218 Feld Ave.
High Point, NC 27264
(800) 334-4107
Manufacturer of sawblades, router bits, and shaper cutters; sells through retailers

HIGHLAND HARDWARE
1045 North Highland Ave. NE
Atlanta, GA 30306
(800) 241-6748
www.highlandhardware.com
Retailer of woodworking tools and accessories, hardware, and supplies; also sells through direct mail

HORTON BRASSES, INC.
P.O. Box 95
Cromwell, CT 06416
(800) 754-9127
www.horton-brasses.com
Manufacturer and retailer of reproduction furniture hardware

JESADA TOOLS
310 Mears Blvd.
Oldsmar, FL 34677
(800) 531-5559
www.jesada.com
Manufacturer and direct-mail retailer of router bits and sawblades

LEE VALLEY TOOLS, LTD.
12 East River St.
P.O. Box 1780
Ogdensburg, NY 13669-6780
(800) 871-8158
www.leevalley.com
Manufacturer and direct-mail retailer of woodworking tools, hardware, finishing supplies, and more

MCFEELY'S SQUARE DRIVE SCREWS
1620 Wythe Rd.
P.O. Box 11169
Lynchburg, VA 24506-1169
(800) 443-7937
www.mcfeelys.com
May have started with just square-drive screws, but the current catalog lists a variety of fasteners, drill and router bits, finishing supplies, and other tools

ROCKLER WOODWORKING AND HARDWARE
4365 Willow Dr.
Medina, MN 55340
(800) 279-4441
www.rockler.com
Retailer of a broad range of tools, hardware, and finishing supplies; also sells through direct mail

TREMONT NAIL
8 Elm St.
Wareham, MA 02571
(800) 842-0560
www.tremontnail.com
Manufacturer (since 1819) and retailer of steel-cut nails and reproduction steel hardware

WOODCRAFT
210 Wood County Industrial Park
P.O. Box 1686
Parkersburg, WV 26102-1686
(800) 225-1153
www.woodcraft.com
Retailer of woodworking tools and accessories, hardware, supplies; also sells through direct mail

METRIC CONVERSION CHART

INCHES	CENTIMETERS	MILLIMETERS	INCHES	CENTIMETERS	MILLIMETERS
⅛	0.3	3	13	33.0	330
¼	0.6	6	14	35.6	356
⅜	1.0	10	15	38.1	381
½	1.3	13	16	40.6	406
⅝	1.6	16	17	43.2	432
¾	1.9	19	18	45.7	457
⅞	2.2	22	19	48.3	483
1	2.5	25	20	50.8	508
1¼	3.2	32	21	53.3	533
1½	3.8	38	22	55.9	559
1¾	4.4	44	23	58.4	584
2	5.1	51	24	61.0	610
2½	6.4	64	25	63.5	635
3	7.6	76	26	66.0	660
3½	8.9	89	27	68.6	686
4	10.2	102	28	71.1	711
4½	11.4	114	29	73.7	737
5	12.7	127	30	76.2	762
6	15.2	152	31	78.7	787
7	17.8	178	32	81.3	813
8	20.3	203	33	83.8	838
9	22.9	229	34	86.4	864
10	25.4	254	35	88.9	889
11	27.9	279	36	91.4	914
12½	30.5	305			